T0141687

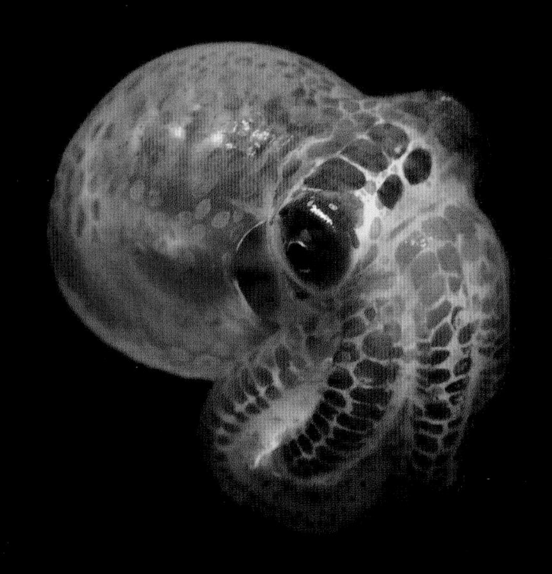

OCTOPUS, SQUID & CUTTLEFISH

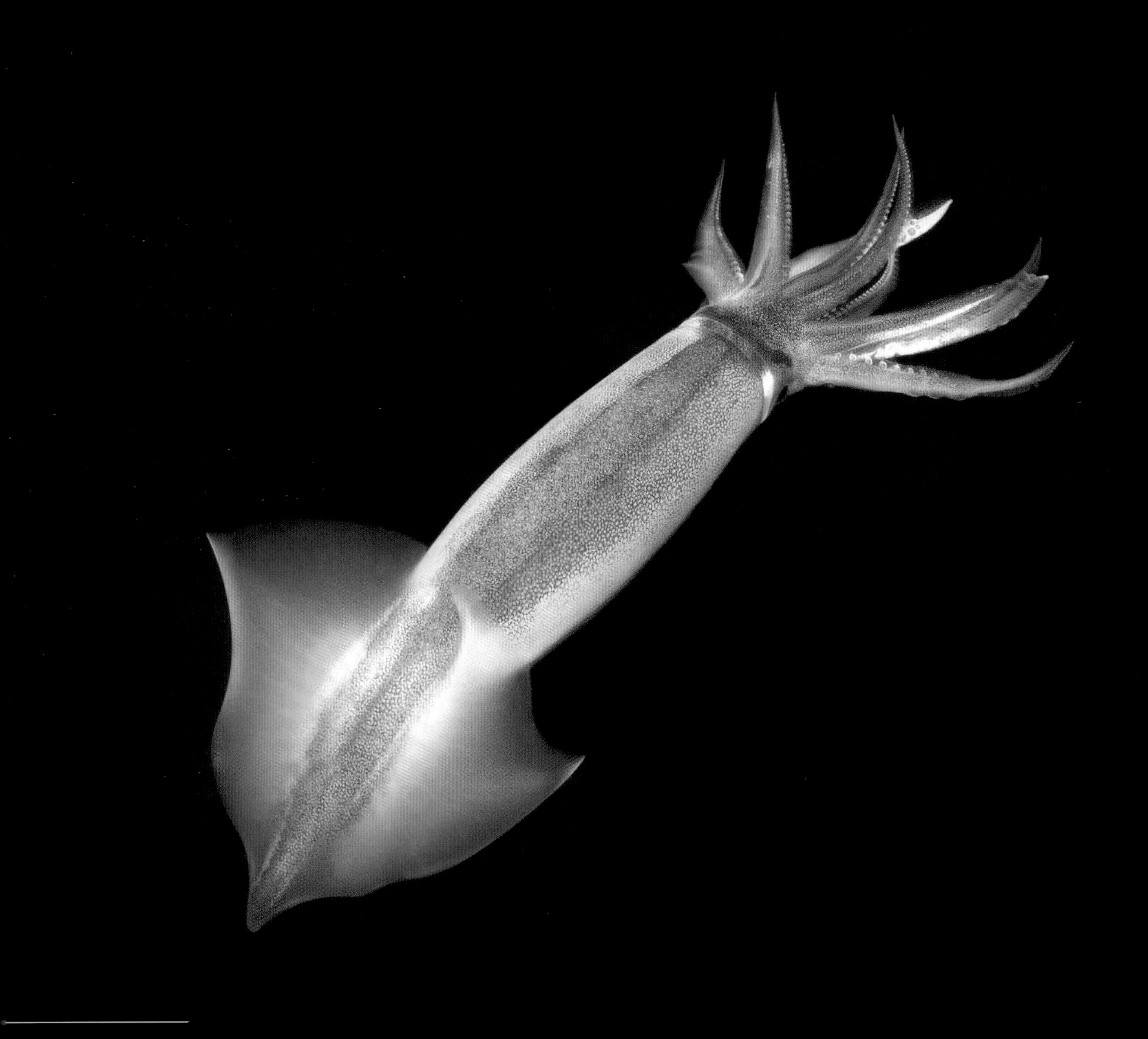

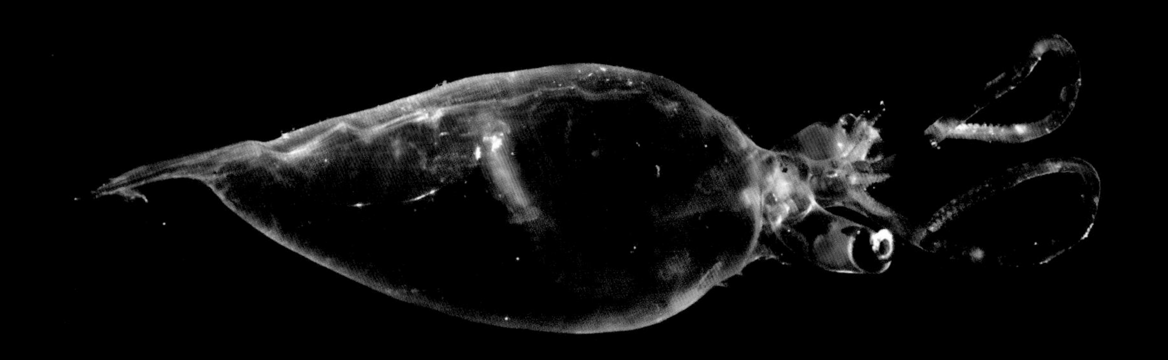

OCTOPUS, SQUID & CUTTLEFISH

A VISUAL, SCIENTIFIC GUIDE TO THE OCEANS' MOST ADVANCED INVERTEBRATES

ROGER HANLON • MIKE VECCHIONE • LOUISE ALLCOCK

THE UNIVERSITY OF CHICAGO PRESS

The University of Chicago Press, Chicago 60637

For more information, contact the University of Chicago Press, 1427 E. 60th St., Chicago, IL 60637.
Published 2018
Printed in Singapore

27 26 25 24 23 22 21   2 3 4 5

ISBN-13: 978-0-226-45956-1 (cloth)
ISBN-13: 978-0-226-45973-8 (e-book)
DOI: https://doi.org/10.7208/chicago/9780226459738.001.0001

Library of Congress Cataloging-in-Publication Data

Names: Hanlon, Roger T., author. | Vecchione, Michael, author. | Allcock, A. Louise, author.
Title: Octopus, squid, and cuttlefish : a visual, scientific guide to the oceans' most advanced invertebrates / Roger Hanlon, Mike Vecchione, and Louise Allcock.
Description: Chicago : The University of Chicago Press, 2018. | Includes bibliographical references and index.
Identifiers: LCCN 2018010067 | ISBN 9780226459561 (cloth) | ISBN 9780226459738 (e-book)
Subjects: LCSH: Octopuses. | Squids. | Cuttlefish.
Classification: LCC QL430.2 .H38 2018 | DDC 594/.56—dc23
LC record available at https://lccn.loc.gov/2018010067

This book was conceived, designed, and produced by
Ivy Press
An imprint of The Quarto Group
The Old Brewery, 6 Blundell Street
London N7 9BH, United Kingdom
T (0)20 7700 6700 F (0)20 7700 8066
www.QuartoKnows.com

Publisher Susan Kelly
Creative Director Michael Whitehead
Editorial Director Tom Kitch
Commissioning Editor Kate Shanahan
Project Editor Joanna Bentley
Designer Kevin Knight
Concept Design Grade Design
Picture Researcher Alison Stevens
Illustrator John Woodcock

Lithocase images Front: Ardea/© Valerie & Ron Taylor.
Back: Nature Picture Library/David Shale.

CONTENTS

INTRODUCING THE CEPHALOPODS

BRAINY, COLORFUL, FAST, SOPHISTICATED, STRANGE, inspiring—cephalopods have been on the planet for about 500 million years and have fascinated humans for thousands of years. Aristotle marveled at their mesmerizing color changes, and they were widely popularized in modern times by Jules Verne, Jacques Cousteau, and Peter Benchley. Scientists have studied them to advance basic and applied sciences, and the oceans depend on them for ecological balance and as a food source for many key predators. They are an evolutionary oddity, having developed the largest and most complex brains among invertebrate animals.

Octopuses, squids, cuttlefishes, and nautiluses are among the most beautiful animals on earth, and they have evolved large brains, keen senses, and complex behaviors that rival those of fishes and birds. They possess a very special quality known as Rapid Adaptive Coloration. That is, most of them can change their skin patterning and overall appearance in a fraction of a second to help deploy a wide range of adaptive behaviors such as camouflage, alarm, threat, predator escape, and mate attraction. To accomplish these visual illusions, they have developed elegant skin pigments and reflectors (or bioluminescent photophores) that produce colorful and jewel-like body patterns—each pattern tuned to a specific function and ocean viewer. These animals are both attractive and enigmatic to the viewing public, yet there are considerable misconceptions about their abilities. This book presents factual and exciting descriptions that highlight their sophisticated behaviors and the key roles that they play in modern oceans, accompanied by stunning photographs of a wide range of species.

WHY A BOOK ON CEPHALOPODS?

Cephalopods are considered a charismatic animal group due to their colorful appearances, complex behaviors, and overall strangeness compared with most other animals. Some are huge (the giant squid) and some are very small and poisonous

(blue-ringed octopus). Cephalopods occupy many oceanic niches and in some cases their numbers are massive—consider that sperm whales feed almost exclusively on squids, and that their biomass alone is estimated by some to equal the total biomass of all human fisheries. Most cephalopod species occupy a central position in ocean food chains, being a primary food source for many marine mammals (whales, dolphins, seals), birds (penguins, petrels), and a vast array of fishes (including swordfishes, groupers, snappers, barracudas, eels). They are an ideal meal because they have few or no hard parts and are almost entirely muscle protein; indeed the cephalopods themselves are the subject of targeted fisheries worldwide. They are totally unlike any other animal group—invertebrate or vertebrate—and are worthy of scientific study in their own right. The public find them fascinating and weird, or simply like to eat them as a basic protein source.

In this age of biotechnology, engineers and biomedical scientists are looking to the natural world for bio-inspired approaches to solve problems of interest to human society, and cephalopods have attributes that society and industry would like to incorporate into modern devices. For example, soft, flexible yet strong robotic arms based on octopus arms

Opposite Octopus camouflage is extraordinarily effective and diverse. They tune the color and pattern to the background and even make their skin bumpy to resemble the 3D texture of the surrounds.

Left Even translucent squids like this one use their dynamically controlled chromatophores to enhance camouflage in the water column.

Below The blue-ringed octopus sometimes expresses its rings weakly for camouflage as in this image, but can also brighten them as a warning signal to approaching predators.

would be welcomed by the medical community as well as search and rescue teams working in collapsed buildings. And who would not like to have the ability to change the color and pattern of their clothes, cars or other objects the way that cephalopods do?

On the other hand, there are some misconceptions about the real capabilities of cephalopods (what kind of "smart" are they?) and we attempt to put these into perspective with known scientific knowledge while simultaneously providing justice to their many fascinating attributes.

WHAT IS A CEPHALOPOD?

It is hard to imagine that an octopus or squid is related rather closely to chitons, snails, and oysters, but they are all in the Phylum Mollusca. Yet the Class Cephalopoda has evolved very differently from other classes of molluscs. Octopuses, squids, cuttlefishes, and nautiluses are invertebrates—that is, they have no backbone. This does not mean that they are "simple" or "primitive" or "lower" animals. As the reader may come to recognize in this book, cephalopods are sophisticated animals with a full array of sense organs, a complex brain, and highly diverse behaviors. Clichés and stereotypes about them abound, with varying degrees of accuracy.

"Marine molluscs on steroids" might be one glib way to describe cephalopods, yet this jocular analogy may be more truth than fiction since cephalopods have hormones that tend to supercharge some of their critical behaviors, such as male–male cuttlefish fights when they compete for a female mate (see chapter 4). "Live fast and die young" is a cliché commonly used to describe cephalopods because they grow very fast (sometimes doubling their body weight every week in young stages) and many of them live only a year or less. Some

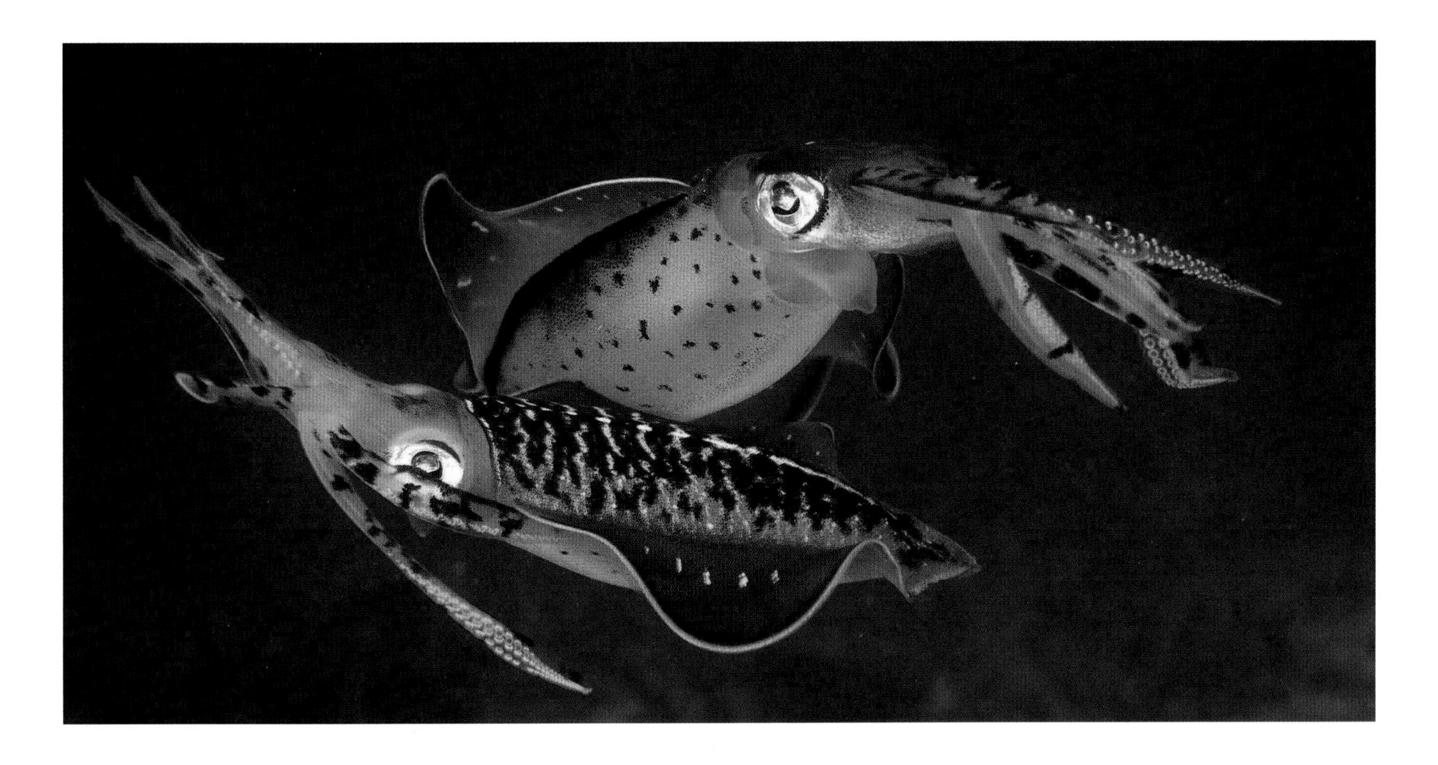

cuttlefishes and squids increase their weight from less than an ounce (a few grams) to 11–22 lb (5–10 kg) in 6–12 months (a few even faster). It takes humans 24 months to grow this large. Cephalopod growth hormones are far more powerful than those in other molluscs. Their extraordinary growth rates are in stark contrast to the slow growth and longer lives of other molluscs such as snails, clams, and oysters. This general life-history strategy is different from other invertebrates too, but is more in line with some groups of marine teleost fishes, which have sophisticated sensory systems, brains, and behaviors. Those fishes, however, typically have longer lives and more standard life histories. In short, cephalopods are just different from any other class of animals.

Squids are often referred to as "jet-powered torpedoes of the sea," and the octopuses and cuttlefishes frequently live up to their popular title of "kings of camouflage." These and other aphorisms are the way that most people view cephalopods. "Red devils" is a local name for the Humboldt squid found off Peru and Mexico because fishermen fear their large size and aggressive behavior (these squids can be as large as small humans, and they cannibalize one another). Some descriptors of cephalopods are not so glamorous: sport fishermen usually think of squids as "bait" because squids are part of the natural diet of game fish such as billfish, grouper, and flounder, and a dead squid on a fishing line is an effective fishing method.

Opposite Cuttlefish can switch from camouflage to conspicuous signaling in one fifth of a second! Here the European cuttlefish is showing a conspicuous zebra pattern.

Above Male squids fight each other with changeable body patterns—a sort of "gentlemanly" duel of visual displays that usually resolves the fight without physical injury.

Right The mimic octopus can shape its body to look very unlike an octopus, and can also release a plume of dark ink to startle and confuse an approaching predator. The very long thin arms are characteristic of sand-dwelling octopuses.

In our zoological view, cephalopods are simply marvels of nature. They took a sharp turn along the tree of life and developed a body form and a way of living that are distinctly different from those of any other animal group. Anyone who has been privileged to watch a squid or octopus on a coral reef knows how mesmerizing it can be, and can immediately appreciate the complex cognitive abilities of these alien-like creatures.

DISTINGUISHING CHARACTERISTICS OF THE FOUR GROUPS

Nautilus appears to be a vestige of the past. It lives in a spiral shell and is more like a slow-swimming snail than a fast, modern squid, octopus, or cuttlefish. It has large but primitive eyes and a relatively small brain compared with modern cephalopods, and it lives exclusively in deep water along coral reef slopes.

The two major groups of modern coleoid cephalopods—the octopodiformes and the decapodiformes—have large brains, complex behavior, large image-forming eyes, an ink sac, fast locomotion, and rapid adaptive coloration. Yet they are very different in body form and habits.

Octopuses have eight long, well-developed arms lined with suckers. Some have no large rigid internal structures and so their bodies have maximum flexibility to enable them to exploit crevices and other tight spots in the bottom habitats in which they live worldwide.

Squids have a long tubular shape and their mantle (the main part of their body) is supported partly by a rigid gladius under the skin. They live in the water column and are the strongest swimmers among cephalopods thanks to their streamlined shape and unusually strong jet propulsion. However, many are negatively buoyant and have to swim constantly to stay in the water column. Squids are the "wide open travelers" of the seas, often swimming many miles or vertically migrating up and down thousands of feet each day.

Cuttlefishes are a bit squid-like in body form but are stouter and, importantly, they have a rigid "cuttlebone" into which

CEPHALOPOD CLASSIFICATION CHART

PHYLUM: **Mollusca**
CLASS: **Cephalopoda**

SUBCLASS: **Nautiloidea**
- FAMILY: Nautilidae

SUBCLASS: **Coleoidea**
DIVISION: **Neocoleoidea**
SUPERORDER: **Octopodiformes**
- ORDER: **Vampyromorpha**
  - FAMILY: Vampyroteuthidae
- ORDER: **Octopoda**
- SUBORDER: **Cirrata**
  - FAMILY: Cirroctopodidae
  - FAMILY: Cirroteuthidae
  - FAMILY: Opisthoteuthidae
  - FAMILY: Stauroteuthidae
- SUBORDER: **Incirrata**
  - SUPERFAMILY: Argonautoidea
    - FAMILY: Alloposidae
    - FAMILY: Argonautidae
    - FAMILY: Ocythoidae
    - FAMILY: Tremoctopodidae
  - SUPERFAMILY: Octopodoidea
    - FAMILY: Amphitretidae
    - FAMILY: Bathypolypodidae
    - FAMILY: Eledonidae
    - FAMILY: Enteroctopodidae
    - FAMILY: Megaleledonidae
    - FAMILY: Octopodidae

SUPERORDER: **Decapodiformes**
- ORDER: **Oegopsida**
  - FAMILY: Brachioteuthidae
  - FAMILY: Cranchiidae
  - FAMILY: Cycloteuthidae
  - FAMILY: Gonatidae
  - FAMILY: Ommastrephidae
  - FAMILY: Onychoteuthidae
  - FAMILY: Thysanoteuthidae
  - Architeuthid families
    - FAMILY: Architeuthidae
    - FAMILY: Neoteuthidae
  - Chiroteuthid families
    - FAMILY: Batoteuthidae
    - FAMILY: Chiroteuthidae
    - FAMILY: Joubiniteuthidae
    - FAMILY: Magnapinnidae
    - FAMILY: Mastigoteuthidae
    - FAMILY: Promachoteuthidae
  - Enoploteuthid families
    - FAMILY: Ancistrocheiridae
    - FAMILY: Enoploteuthidae
    - FAMILY: Lycoteuthidae
    - FAMILY: Pyroteuthidae
  - Histioteuthid families
    - FAMILY: Histioteuthidae
    - FAMILY: Psychroteuthidae
  - Lepidoteuthid families
    - FAMILY: Lepidoteuthidae
    - FAMILY: Octopoteuthidae
    - FAMILY: Pholidoteuthidae
- ORDER: **Myopsida**
  - FAMILY: Australiteuthidae
  - FAMILY: Loliginidae
- ORDER: **Sepiida**
  - FAMILY: Sepiidae
- ORDER: **Sepiolida**
  - FAMILY: Sepiadariidae
  - FAMILY: Sepiolidae
- ORDER: **Spirulida**
  - FAMILY: Spirulidae
- ORDER: **Idiosepiida**
  - FAMILY: Idiosepiidae
- ORDER UNCERTAIN
  - SUPERFAMILY: Bathyteuthoidea
    - FAMILY: Bathyteuthidae
    - FAMILY: Chtenopterygidae

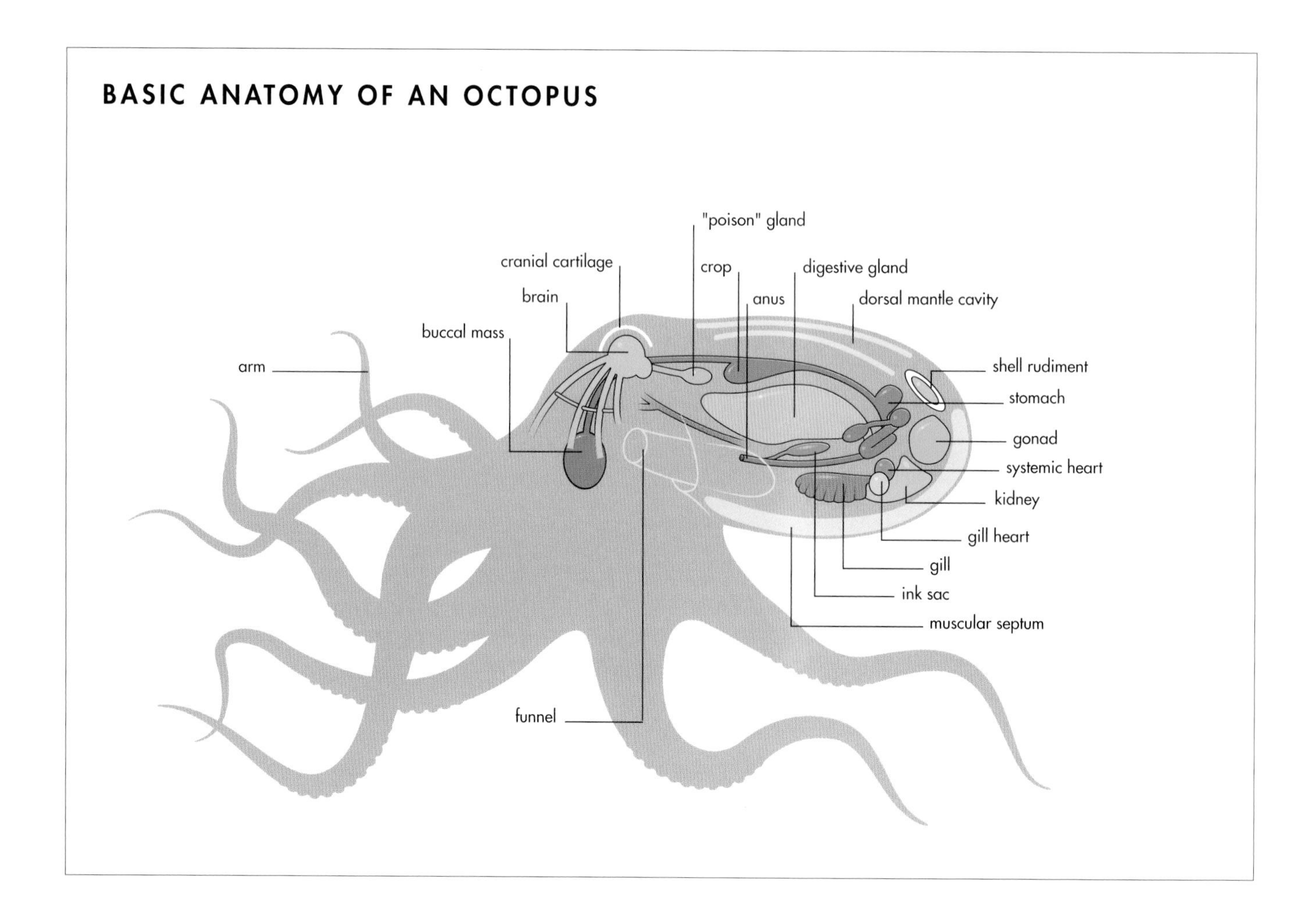

they can pump gas to regulate their buoyancy, so they can hover at any depth to save energy. They tend to live near or on the bottom. Sepiolids resemble cuttlefish but have no cuttlebone and are exclusively bottom dwellers that can bury themselves in the sandy substrates.

BASIC ANATOMY

"Cephalo pod" translates as *"head foot"* and indeed this body form, with the arms and tentacles attached directly to the head, is very distinctive and quite unusual among animals. The "head" has two large image-forming eyes, one on either side, and between them is the large brain that may have more than 30 distinct lobes and 150 million tiny nerve cells. The mouth is at the confluence of the eight arms, it has a strong parrot-like beak and a rasping tongue (the radula) to break food into small bits because the esophagus passes right through the middle of the brain to the body. The body and the typical organs such as stomach, kidney, heart, gills, and reproductive organs are in the "mantle" of the animal. In most cephalopods, water is drawn into the mantle cavity and then forcefully expelled out of the funnel to provide typical breathing or jet propulsion. If the funnel is directed anteriorly the cephalopod jets backwards, and if pointed posteriorly the animal can jet forward.

The skin of cephalopods is especially unusual and well developed with thousands (sometimes millions) of chromatophore organs that can open or close in a fraction of a second to produce colorful patterns in the skin for camouflage and communication. Octopuses and cuttlefishes have an additional feature that is unique in the animal kingdom: specialized bumps in the skin called papillae that

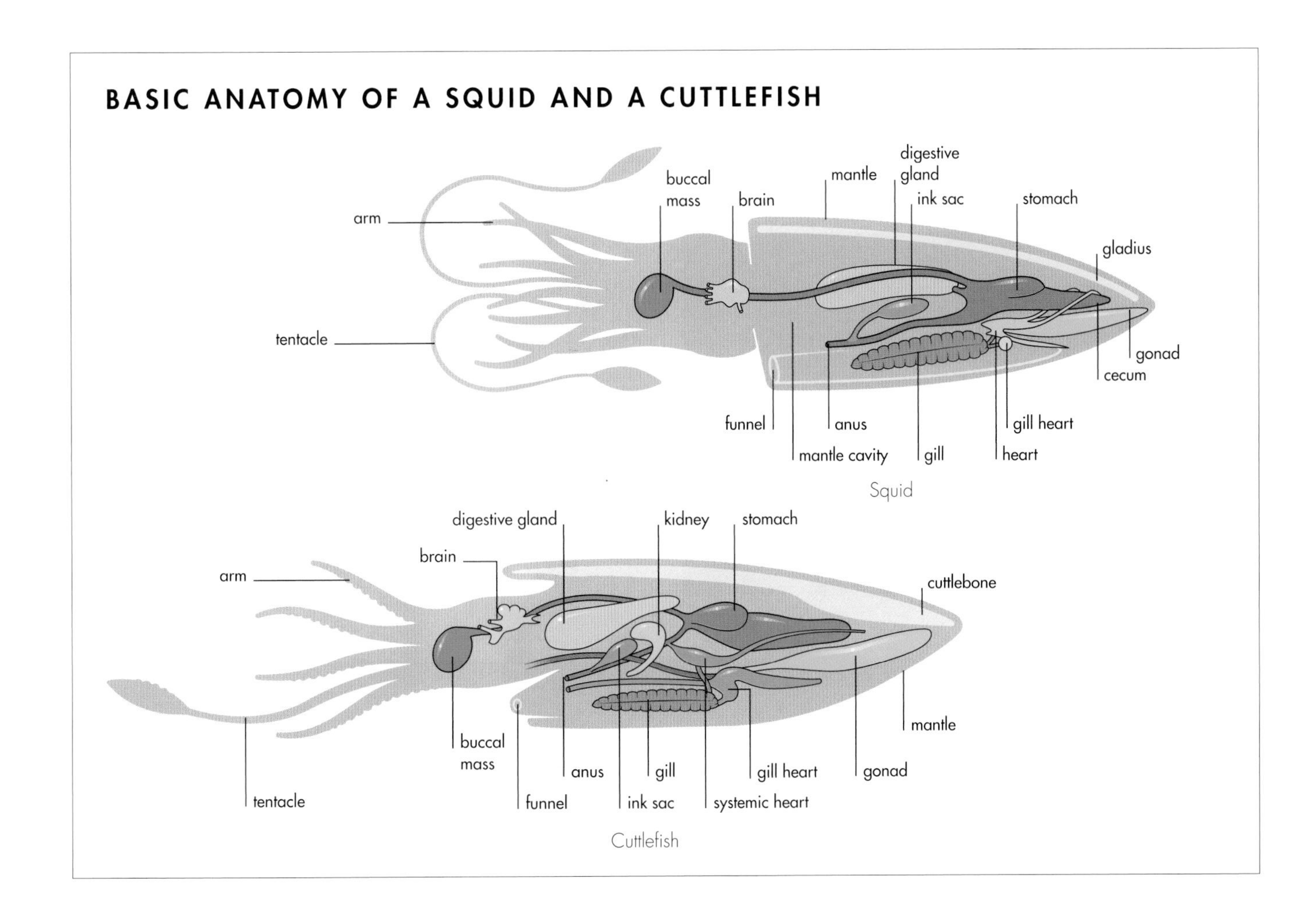

can be fully extended for camouflage, or fully retracted for smooth skin for swimming and jetting. Suckers on the arms and tentacles are another hallmark of cephalopod anatomy. They are not only grasping organs but those on the arms also contain sensory cells for touch and taste.

WHERE DO THEY LIVE?

Cephalopods live in all of the world's oceans from the Arctic to the Antarctic. Throughout those oceans they occupy all habitats from the warmest tide pools in the tropics to very cold depths at 16,500 ft (5,000 m) down in the open ocean, and even under the ice in polar seas.

No cephalopods occur in fresh water, and only a few can tolerate salinities slightly less salty than oceanic water. Small squids of the genus *Lolliguncula* have evolved the capability to occupy water with lower salinity (about half as salty as ocean water). *Lolliguncula* can slowly conform to this less salty water over the course of a few hours.

In the surface waters of the open ocean, schooling squids are distributed over enormous areas and are difficult to study or

Opposite and above The basic anatomy of cephalopods. Each has a mantle (body sac) that contains all of the typical internal organs. The eyes and brain are in the "head," which in turn is connected to the arms. This strange "head-foot" body plan is why they are called "cephalo-pods." All eight arms are lined with suckers, which provide touch and smell capability as well as suction for fighting or prey retention. Squids and cuttlefishes also have two long tentacles that are usually retracted and hidden, but can be shot out very fast to grab prey and pull them back for consumption.

Left *Lolliguncula brevis* is a species that occurs in shallow, low-salinity areas, usually bays and estuaries.

Below The day octopus of the Indo-Pacific is a king of camouflage, and has mastery of its skin papillae—bumps of different shapes that can be extended to different lengths or completely flattened to produce smooth skin to enable fast swimming.

Opposite The firefly squid can produce blue bioluminescence, as seen in very shallow water on this beach in Japan. The squids seem to use this visual signaling to one another during reproductive behavior.

even to capture as fisheries targets. Many open ocean squids stay in the dark on purpose by remaining below about 900 ft (275 m) during the day and vertically migrating near the surface during the night to follow food sources such as shrimps and lantern fishes. These squid species have visual systems that are well adapted to darkness and to bioluminescence. Other cephalopods live throughout the deep sea where sunlight never penetrates. In these depths—from 1,640 to 16,400 ft (500 to 5,000 m)—the cephalopods are a major part of the fauna and their body forms are particularly peculiar and diverse.

It is difficult to find cephalopods on shallow coral reefs, kelp forests, sand plains, and seagrass meadows because their camouflage is so effective. Yet they are common in those habitats, although some are strictly nocturnal, some are day active, and others are active only during dawn and dusk. For example, the day-active *Octopus vulgaris* in the Caribbean is common in and around coral reefs but divers seldom see them due to the octopus's superb camouflage and stealth behavior as it forages.

ABOUT THIS BOOK

THIS BOOK IS ORGANIZED INTO FIVE CHAPTERS that cover many aspects of the biology, ecology, and commercialization of cephalopods. Each chapter is organized to highlight common as well as unique features that enable cephalopods to thrive throughout the world's oceans. Key concepts that characterize this group of marine animals are provided as well as profiles of representative cephalopod species to help visualize some of their anatomical and behavioral attributes.

CHAPTER THEMES

Chapter One introduces the reader to the form and function of the major body parts—both exterior and interior. The diverse sensory systems for vision, taste, touch, and smell are described, as are the various body parts that enable crawling, swimming, jet propulsion, color change, and reproduction. Chapter Two takes a broad view of the evolutionary history of cephalopods and their relationships with other major animal groups. The fossil record for ancient cephalopods is very rich and shows their deep history all the way back to the Cambrian era, yet the modern cephalopods with their soft bodies have left much less fossil evidence, which obscures their phylogenetic relationships. Chapter Three focuses on the peculiar lifestyles of cephalopods that enable them to compete directly with fishes in all marine habitats. Chapter Four takes a more introspective approach and sorts out a wide range of complex behaviors that are enabled by sophisticated sensory and motor systems coordinated by a very large brain. This chapter also touches on more difficult and speculative subjects such as learning and memory, tool use, play, and intelligence. Chapter Five informs the reader of various ways in which humans benefit from cephalopods, either consuming them as a food source or using them to inspire new developments in biotechnology and medicine.

SPECIES PROFILES AND CEPHALOPOD IMAGERY

Each chapter is enriched by species accounts that inform the reader of the rich diversity of special attributes that

Opposite left Suckers are complex, multifunctional organs that are richly innervated for touch and smell, and to create suction for crawling, fighting, tasting, touching, and grasping.

Opposite right A small shoal of *Sepioteuthis sepioidea*, near a coral reef in the Cayman Islands.

Above The Humboldt squid is very large, fast, and aggressive. It ranges widely along the Pacific coasts from Chile to Alaska and is a major oceanic predator.

cephalopods have evolved over millennia. Each profile highlights some of the unique and peculiar biological traits of that individual species, as well as common names, sizes, habitats, feeding habits, and key behaviors.

Many photographs and drawings have been chosen with the aim of providing the reader with images with as much biological richness as possible. Cephalopods are very active animals and their complex behaviors are difficult to capture in still images. However, the beauty of their skin displays and the various interactions they have with predators, prey, and members of their own species can often be captured in still photographs. Fortunately the cephalopods are popular among amateur and professional photographers as well as biologists, so we were able to assemble a repertoire of imagery that represents features that are distinctive to this successful group of marine invertebrates.

CEPHALOPOD ANATOMY

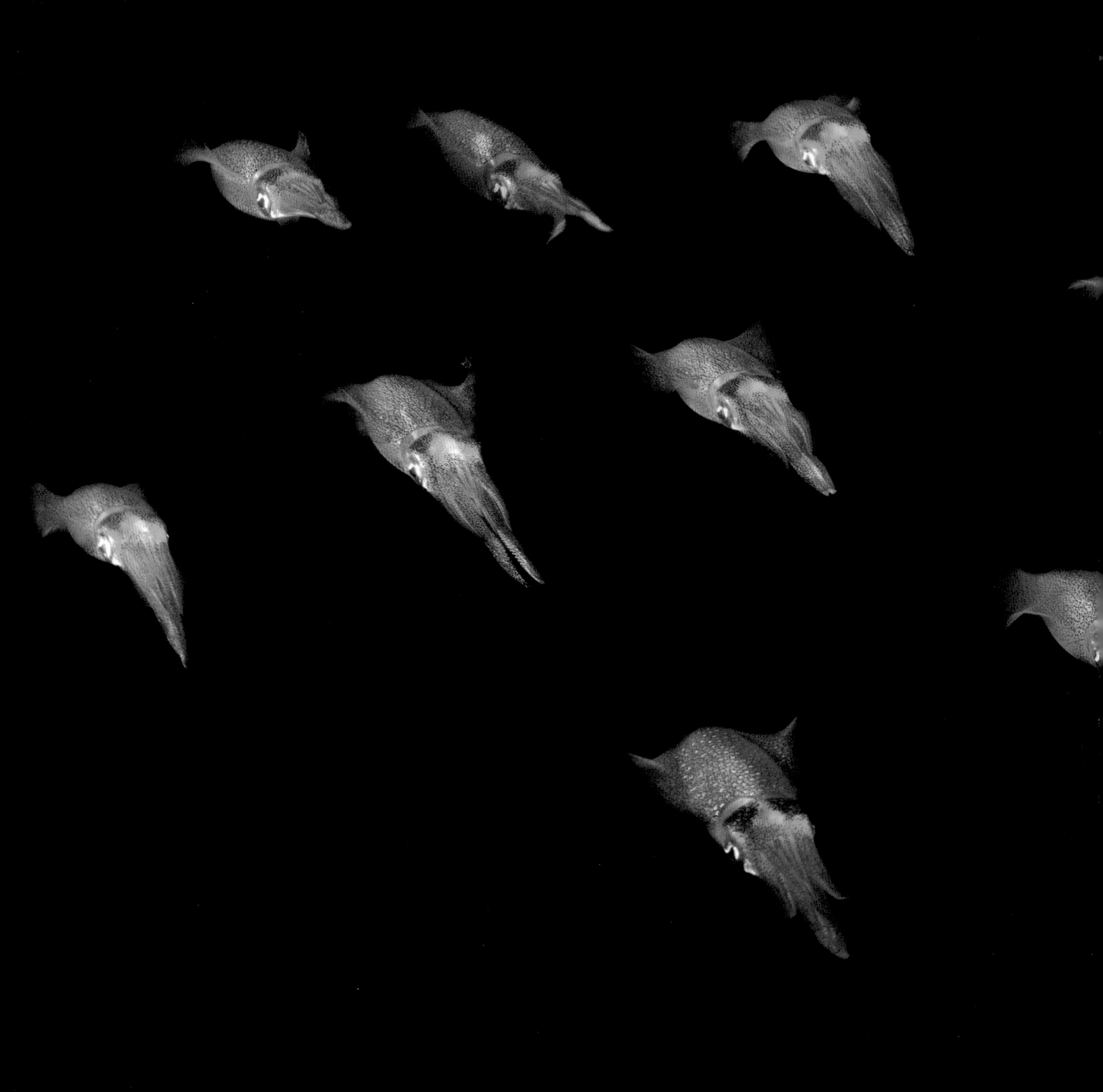

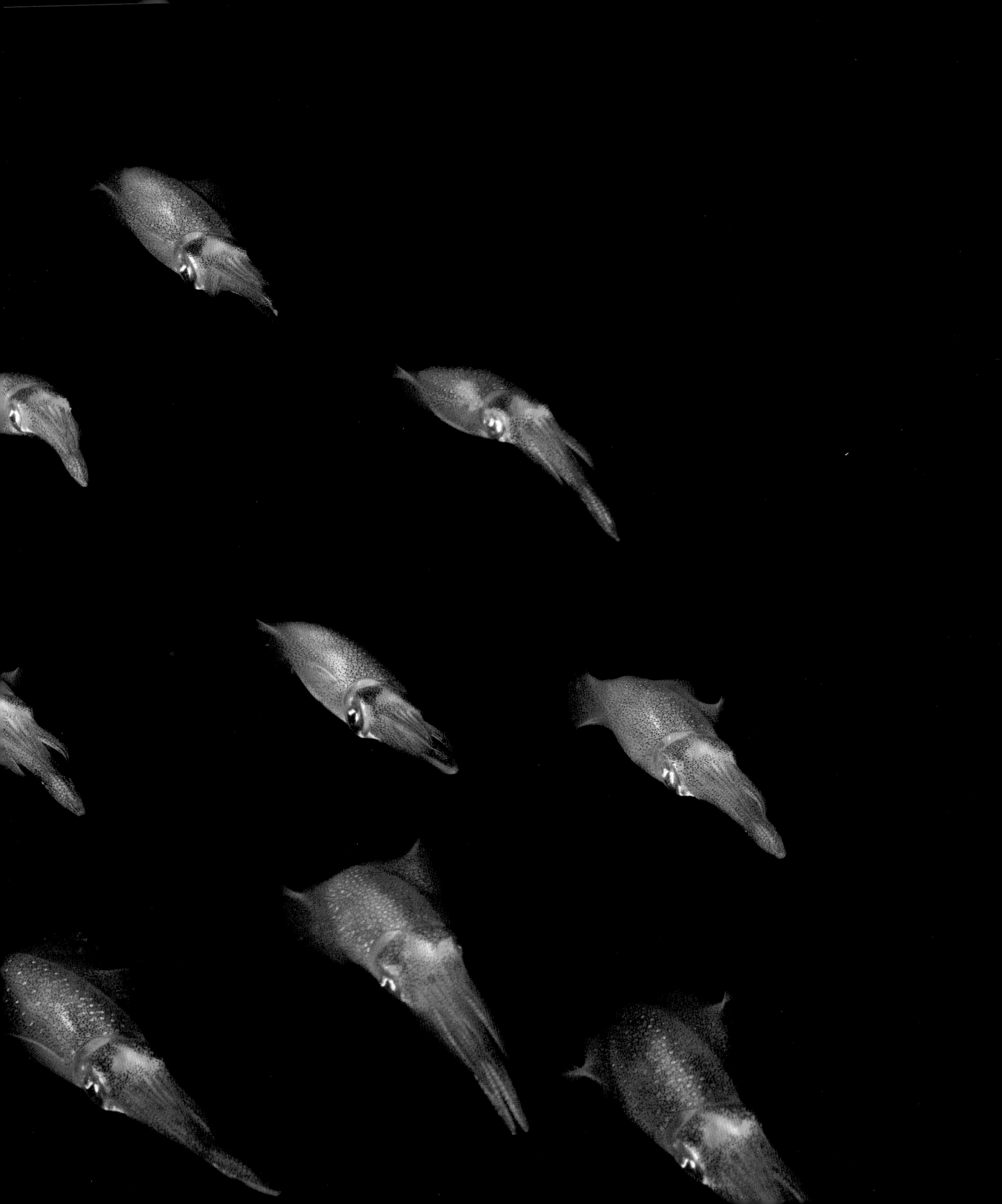

ADVANCED INVERTEBRATES

COMPLEX, HIGHLY DEVELOPED NERVOUS AND sensory systems are typical features of living cephalopods. Especially noteworthy are the image-forming eyes and the complex brain, which has evolved from the nerve ring around the esophagus of other molluscs. The ability of most cephalopods to change their appearance rapidly because of specialized pigment organs and reflector cells in the skin that are under direct nervous control makes observation of living animals fascinating. Other noteworthy peculiarities, such as movement by muscles squeezing other muscles, are also characteristic of cephalopods.

EXTERNAL STRUCTURE

Three regions can easily be discerned on a cephalopod. From anterior to posterior, these regions are:

1) the arms and tentacles surrounding the mouth
2) the head, on which the eyes are prominent
3) the sac-like mantle, or body, to which fins may be attached.

This overall structure is less distinct in nautilids, which have external shells into which they can withdraw, but is still recognizable.

THE CROWN OF ARMS

Cephalopods other than nautilids have either eight or ten arms surrounding the mouth. On those with ten, two arms are modified into either tentacles (decapods) or filaments (vampires). The arrangement of the arm crown is different in nautilids from that in all other living cephalopods. Nautilids have approximately 90 arms arranged in two rings around the mouth. Each arm is retractable within an outer sheath.

The oral surface of each arm is covered by adhesive transverse ridges, rather than by the suckers typical of other cephalopods, which have either eight or ten arms surrounding the mouth. Hence, all living cephalopods other than nautilids have eight arms and some additionally have two tentacles or two filaments.

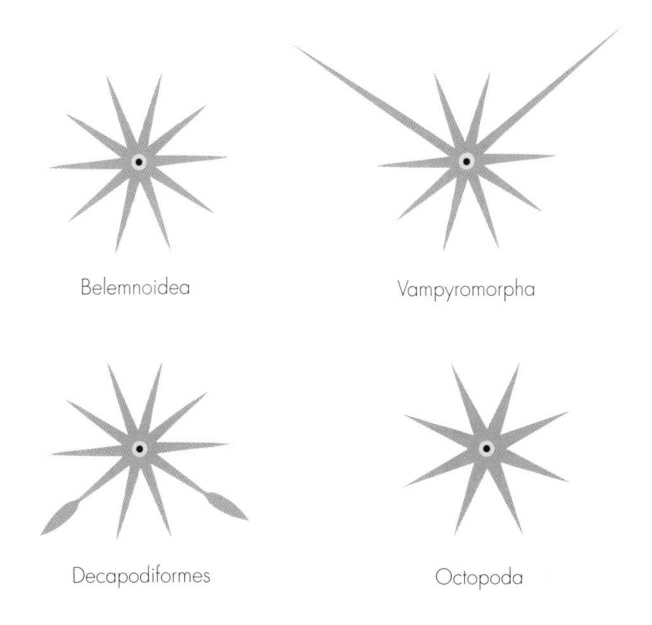

TENTACLES

The most obvious difference between arms and tentacles on a decapod is in the arrangement of armature (suckers, some of which may be modified into hooks). The arms of most cephalopods have one, two, or four longitudinal series of suckers or hooks extending along the entire oral surface. Tentacles tend to be longer than arms and to have the armature concentrated toward the far end, which is often expanded to form a tentacular club. The filaments of vampires are very long and thin with no armature, and can be retracted into pockets.

Although both decapods and vampires have eight arms and two modified appendages, those modifications are not of the same two appendages. In decapods, the tentacles are ventrolateral, whereas in vampires, the filaments are dorsolateral. Although many people think that the two arms that have been lost in the octopods, out of the general ancestral pattern of ten arms, are the tentacles as found in decapods, it is more likely that it was the dorsolateral arms that are modified in vampires that were lost in octopods.

Opposite The head and arm crown of a Humboldt squid, *Dosidicus gigas*.

Top right Major differences in arm crowns among coleoids. The extinct Belemnoidea exhibits the ancestral condition of ten equal, undifferentiated arms. In Vampyromorpha the dorsolateral pair is modified whereas in Decapodiformes the modified tentacles are ventrolateral.

Middle right The tentacular club of a glass squid (family Cranchiidae). Suckers are concentrated on the expanded distal end of the tentacle.

Bottom right Dorsal view of the pelagic oceanic bobtail *Heteroteuthis dispar*, showing the typical structure of mantle (body) with fins, head with eyes, and arm crown.

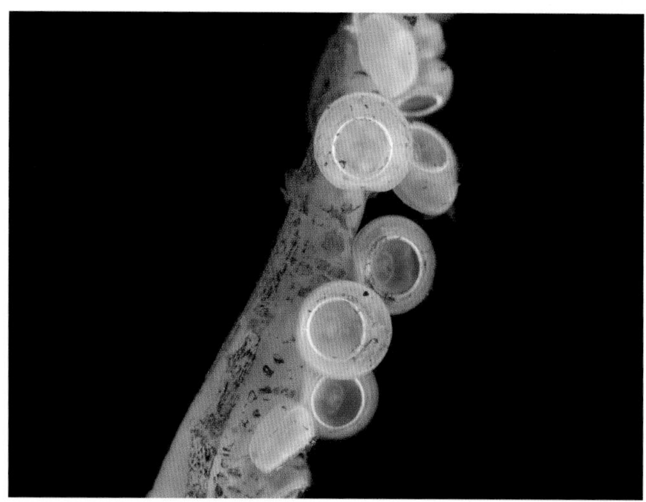

SUCKERS

Suckers are of exceptional use to cephalopods—the arms and tentacles essentially have the task of delivering the suckers to a destination to conduct feeding, tasting, attachment, or movement. Basic differences exist in the structure of suckers among the major groups. Octopod suckers are radially symmetrical with cylindrical bases that are either broader than the suction cup or only slightly constricted. Decapod suckers are bilaterally symmetrical and their bases are narrow stalks. Therefore, an octopod sucker may be described as looking like a volcano whereas a decapod sucker looks a little like a ball on a stick.

The surface around the inner opening of an octopod sucker has a thin cuticular lining, while the opening of a decapod sucker is characterized by a rigid chitin-like ring that digs into captured prey upon suction and resists the shear forces caused by a wriggling fish or shrimp. These rings may be smooth, notched (referred to as "blunt teeth"), or serrated ("sharp teeth"). The tooth-like structures are of unequal sizes around the opening of the sucker. The result can be elongated sharp teeth in one area of the sucker ring and either blunt teeth or a smooth region on the opposite area. In several squid families, the central elongated tooth in some suckers becomes hyper-developed, forming a hook reminiscent of a cat's claw.

Top left Biserial suckers on an *Octopus vulgaris* arm. Unlike in decapods the suckers are not stalked.

Middle left Stalked decapod suckers on *Teuthowenia megalops*.

Bottom left Close-up of the suckers of *Dosidicus gigas* showing sucker rings with tooth-like serration.

Opposite A veined octopus, *Amphioctopus marginatus*, flaring the webs between its arms.

Conical muscular structures, like little fingers, are associated with the bases of the suckers in cirrate octopods, vampires, and most decapods. In cirrates and vampires, both of which have suckers in a single series, these structures are called cirri (singular = cirrus) and are found in pairs alternating with the suckers along part, or all, of the arm. The similar structures on decapods are called trabeculae (singular = trabecula) and extend from the outer bases of the outer series of suckers on the arm or tentacle. Cirri and trabeculae very likely evolved from the same ancestral structure.

In many decapods some suckers in the proximal part of the tentacular clubs develop as simple knobs, alternating with normal suckers. When the two clubs are held together, the knobs on one tentacle align with the matching sucker on the other tentacle and each sucker in that part of the club can attach to a knob. Hence, the tentacular clubs can be attached together at their bases by this overall structure known as a tentacular locking apparatus.

WEBS

In addition to the suckers, the arms have a variety of external structures. Membranes often border the oral surfaces of the arms. Because their function appears to be protection of the suckers, these structures are often described as protective membranes. When these membranes are continuous between adjacent arms, they are called webs. The webs are quite deep in some species, essentially connecting the entire arm to its neighbor. Webs in some species can be very thick and fleshy.

On many decapods, the trabeculae are embedded in the protective membranes, in which case they may be referred to as trabecular membranes. Decapods also have a ridge on the surface of the arms and tentacular clubs away from the suckers. These ridges, known as keels, provide hydrodynamic lift to the head/arm end of the animal when swimming tail-first. The keels on the ventral arms of some squids (such as chiroteuthids and mastigoteuthids) are shifted in position to form a groove in which the thin tentacles are held.

THE MOUTH IN THE MIDDLE

Decapods have a structure around the mouth somewhat similar to an inner ring of arms. A membrane around the mouth, the buccal membrane, is supported by six to eight muscular pillars called lappets. The lappets are like little arms; in some taxa (including some cuttlefishes and inshore squid species, and the family Bathyteuthidae) the lappets have tiny suckers like miniature arm suckers. All functions of the buccal membrane are not known, but in some species sperm packages are implanted on the buccal membrane of the female during mating.

In all cephalopods the mouth is surrounded by a ring of fleshy tissue known as the lips. However, the most prominent feature of the mouth is the hard beak, actually a pair consisting of upper and lower beaks. The upper beak is generally more pointed and the lower is broader, but the edges of both are sharp and come together as an efficient cutting device. When the beaks are closed the upper fits within the lower. The scene in the Walt Disney movie of *20,000 Leagues Under the Sea* in which the giant squid attacks the submarine is noteworthy not only for frightening generations of small children, but also because the squid's beak is upside down, looking more like a parrot's beak. The beaks are necessary to cut food into pieces small enough to pass along the esophagus and through the brain. In some species the beaks are also used to create puncture wounds through which to inject poison from the salivary glands into the prey.

Behind the beaks, and between them when the beaks are open, is a tongue-like structure covered with teeth, the radula, a typically molluscan feature. The radula is a ribbon made up of many rows of teeth. It moves forward and backward along a muscular tongue. As each row of teeth crosses the tip of the tongue the teeth become erect to snag pieces of the prey, which are then drawn deeper into the mouth. Below the radula is a salivary papilla used by octopods to dissolve a hole through heavy shells of molluscs and crabs for injection of poison. The beaks, radula, and tongue are embedded within a discrete spherical mass of muscle known as the buccal mass.

Left The mouth, including a pair of hard sharp beaks surrounded by fleshy lips, is in the center of the arm crown of a cephalopod such as this Humboldt squid (*Dosidicus gigas*).

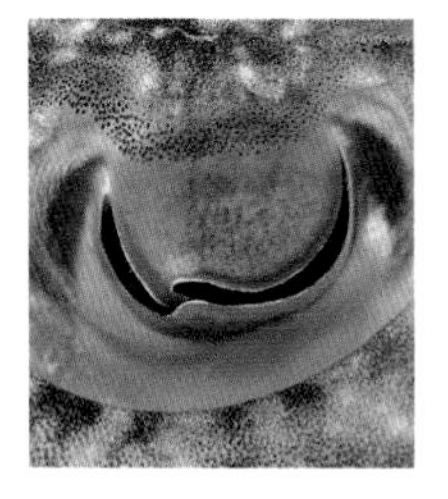

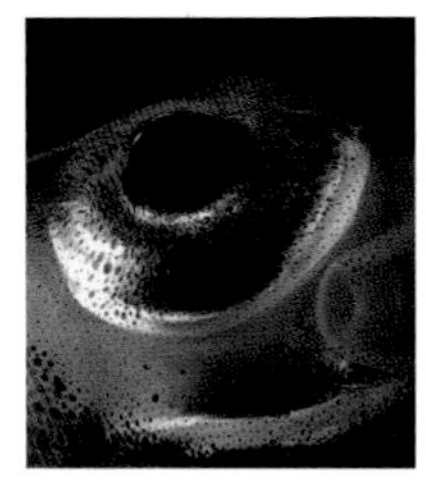

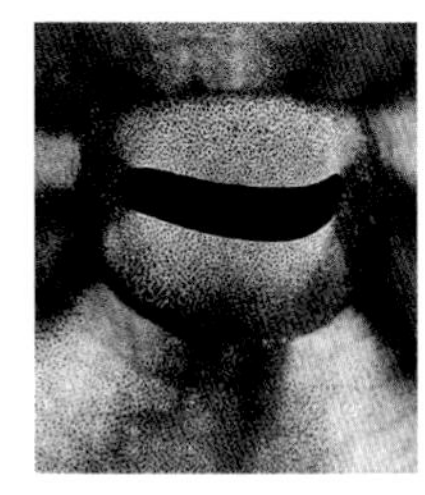

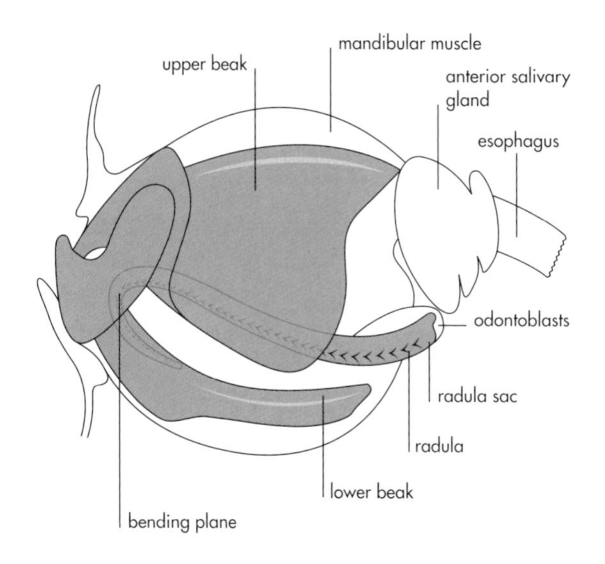

Top left The "pinhole camera" eye of a nautilid.

Left From left to right, the eyes of a cuttlefish, an inshore squid, and an octopod, showing the differences in pupil shape.

Above The buccal mass of an octopod showing the positions of the pair of beaks, along with the radula and the anterior salivary gland.

EYES THAT SEEM FAMILIAR

The eyes are located laterally on the head. Nautilids have simple eyes in which light refracts through a very small hole, open to the sea, to focus on the retina, much like a pinhole camera. Eyes of almost all other cephalopods are "image forming" and have a transparent spherical lens that more efficiently focuses larger amounts of light on the retina. In oceanic squids, the lenses are exposed to seawater. The groups found in coastal or benthic habitats (inshore squids, cuttlefishes, and incirrate octopods) have a transparent corneal membrane covering the lens, presumably for protection. The details of these corneas differ among groups, indicating evolutionary convergence.

The number of photoreceptors in the retina varies from about 4 million in nautiluses through approximately 20 million in octopuses, to estimates of a billion in the eyes of large squids. Although the eyes of neocoleoids seem to be similar to those of vertebrates, a major difference is the orientation of the photoreceptors in the retina. Whereas retinal nerves in vertebrates are at the end of the cell toward the lens (and therefore the light source), those of cephalopods are at the end of the cell away from the lens. The way the photoreceptors are shaped and packed into the retina allows cephalopods to detect polarization of light, which may be important for communication and coordination with others of their own species and for detection of prey. With the exception of one species, the retinas of cephalopods investigated to date have a single pigment, meaning they are "monochromats" and therefore color-blind. However, research has indicated that the strangely shaped pupils of cuttlefishes (W-shaped) and shallow-water octopods (elongate) may allow them to detect colors based on distortion of the image by the edge of the pupil.

Olfactory papillae are also located on the head, their position varying among groups. Although these organs have been presumed to be sensory based on histology, a physiological response to trace chemicals (odors) in seawater has only recently been demonstrated.

WHERE HEAD MEETS BODY

The junction of the head with the mantle varies substantially among the major groups. In most decapods the head is not fused to the mantle but articulates with it at three points. Where the head and mantle meet dorsally in most decapods, a cartilage-like ridge on the head is matched to a similar groove on the interior of the mantle edge. In some squids this ridge and groove system becomes permanently fused. The dorsal mantle edge is actually attached by muscles and skin to the head in some decapods (some bobtails, for example), vampires, and all octopods. Various crests and folds on the dorsal and lateral areas of the posterior head are characteristic of many decapods.

The funnel lies in the funnel groove, a depression in the posterior ventral surface of the head. The funnel is a tube-like structure through which water is expelled from the mantle cavity along with wastes and, when the animal feels the need, ink. As in so many other particulars, the structure of this feature differs in nautilids from that of other cephalopods and has a different name. The funnel of a nautilid, called a hyponome, consists of two muscular flaps that overlap ventrally but are not fused together. Embryological development of the funnel in neocoleoids begins also as two separate flaps that fuse into a tube before hatching. The funnel is flared at the posterior end, which is located within the mantle cavity. The flaring is continuous with a flap of tissue, the collar, extending laterally and dorsally around the head for the entire mantle opening. The collar allows water to be drawn into the mantle around the mantle opening. It closes like a valve, sealing the mantle opening when the mantle contracts, increasing internal water pressure to accelerate expelled water through the funnel. In some decapods, another flap (the funnel valve) is found on the interior of the narrow outer end of the

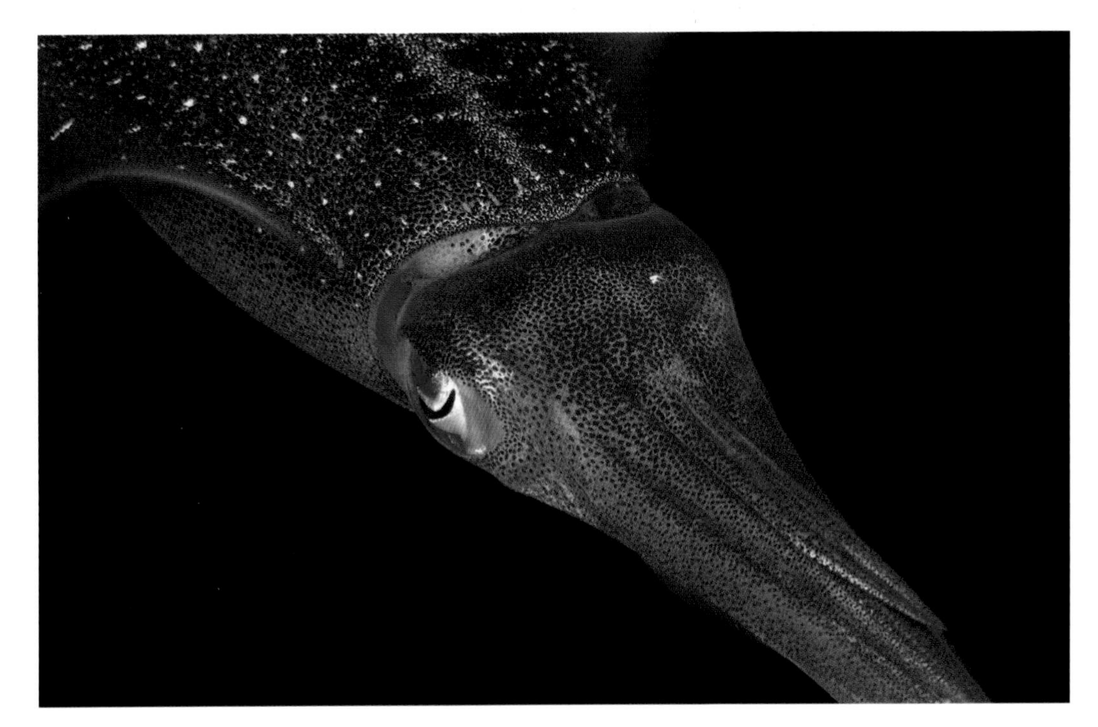

Left Dorsal view of the head and mantle of a Caribbean reef squid, *Sepioteuthis sepioidea*. Beneath the edge of the mantle are the ridge and groove of the dorsal locking apparatus.

Opposite A Pharaoh cuttlefish, *Sepia pharaonis*, with its funnel pointed toward the viewer.

funnel; it prevents water from entering the funnel “the wrong way” when the mantle is expanding and refilling with water. Also inside the funnel is a secretory apparatus called the funnel organ; this produces mucous to form ink blobs and to expel wastes.

In addition to the nuchal cartilage in decapods, there are two funnel-locking cartilages. These structures lock the free edge of the mantle to the funnel to increase the efficiency of mantle contractions. Some locking apparatuses with complex shapes form a very strong bond whereas others are simple and straight, allowing the components to slide in and out, permitting head retraction into the mantle cavity, while preventing slippage sideways. In some species the locking apparatuses are completely fused.

THE MANTLE OR "BODY"

The muscular molluscan mantle is much more than a sac enclosing the viscera. A complex system of muscles and connective tissue interacts to control the volume of the mantle cavity, the water-filled space inside the mantle. Contraction of transverse muscles causes the mantle wall to become thin and to increase in area. This increase in wall area results in the mantle expanding, drawing water into the mantle cavity through the mantle opening and past the collar. Contraction of the circular muscles decreases the volume of the mantle cavity, resulting in an increase in internal water pressure, sealing the collar and expelling water through the funnel. A third set of muscles, oriented longitudinally, prevents the mantle from elongating too much during mantle contraction. Obliquely arranged fibers of connective tissue, together with inner and outer sheaths of connective tissue (the tunics), store energy when they are stretched by the muscular actions outlined above. This stored energy aids the muscles in the cycle of expansion and contraction.

Below A Giant Pacific octopus, *Enteroctopus dofleini*, using its arms to crawl across the sea bottom.

FINS FOR SOME

A pair of fins attaches laterally to the mantle in all decapods, and in vampires and cirrate octopods. Whereas jetting can result in rapid acceleration, swimming with fins uses energy more efficiently. Having both allows dual-mode swimming. The structure of the fins differs among the groups that have fins and the details of fin attachment vary among decapod families. Different types of fin can be used for flapping and undulation as well as for steering.

ELABORATE AND BEAUTIFUL SKIN

A hallmark of cephalopods is their remarkable skin that produces a vast array of appearances for communication and camouflage. The rapid changeability arises from the brain's direct neural control of millions of pigmented chromatophore organs (each one being either yellow, red, or brown) and iridescent cells (producing all colors) in the skin. Octopuses and cuttlefishes tend to have the most complex skin—equivalent, for example, to a high-resolution digital screen—with a dense concentration of small chromatophores.

Squids generally have fewer chromatophores per mm^2 and they are large, hence they are more similar to a low-resolution digital screen. But both are equally fast in their changeability. This skin has an abstract beauty all its own, which will be looked at in more detail in Chapter 4.

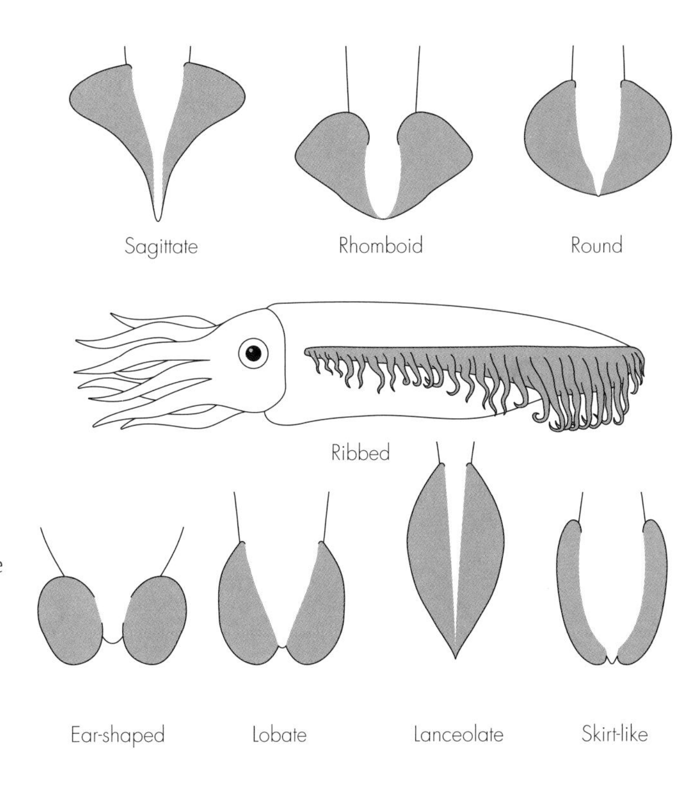

Above Fins of different cephalopod families come in a variety of configurations, each of which functions slightly differently for swimming.

Below Two photos of the same cuttlefish showing chromatophores expanded (left) and contracted (right).

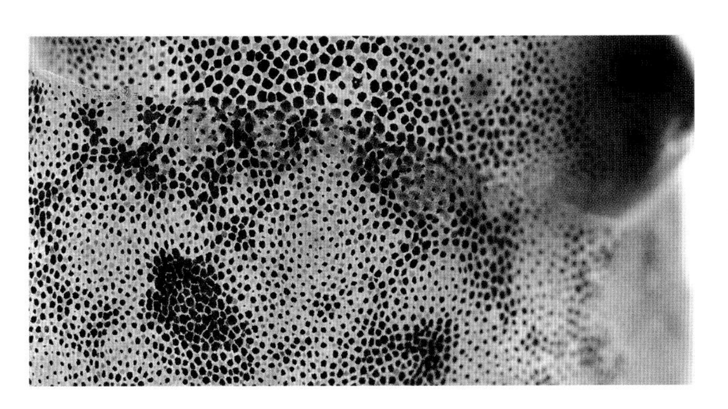

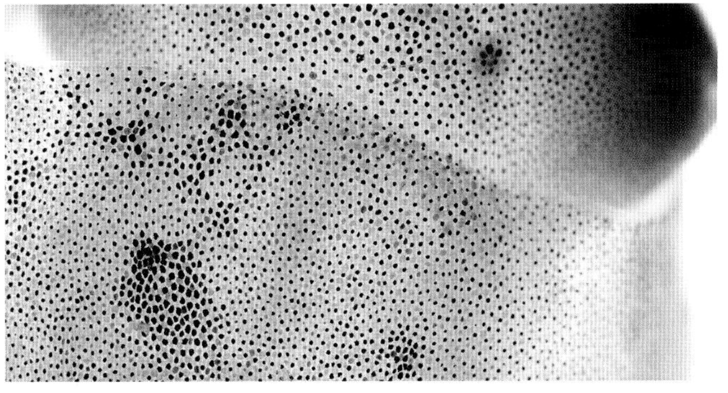

INSIDE THE CEPHALOPOD

CEPHALOPODS SEEM VERY DIFFERENT FROM their closest relatives, the gastropods and other molluscs. Their internal organization, however, is constrained by their evolution from a molluscan ancestor. Only upon detailed examination do the anatomical and physiological characteristics become obvious as modifications of the basic molluscan body organization.

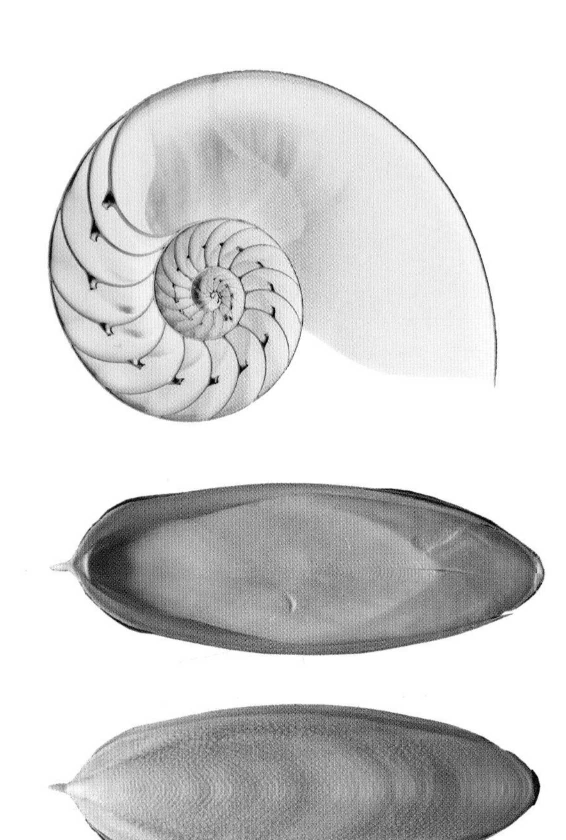

SHELLS AND SKULL

Most cephalopods are thought of as being shell-less, but this is not entirely true. As molluscs, the ancestral condition for cephalopods is the presence of an external shell made up of calcium carbonate, as is found in nautilids. In other living cephalopods with a shell, it is internal, reduced, and often not calcareous in composition. These internal structures, which differ in form among major taxonomic groups, are secreted by a shell sac and are the evolutionary equivalent of the shells of other molluscs. Although internal, the shell sac is actually ectoderm that has formed an internal space during embryonic development. The feature of cephalopod shells that is unique among molluscs is a series of gas-filled chambers termed a phragmocone, well known from fossils and in living nautilids.

Cuttlefish shells retain the phragmocone, as do the peculiar ram's horn squids. Most decapods have a pen-like gladius. Vampires have a similar structure, which is one reason that they were once considered to be more closely related to decapods than to octopods, although this now is believed not to be true. Cirrate octopods have a cartilaginous shell that supports their fins, whereas the shells of incirrate octopods are reduced to a couple of rods that support muscle attachment. Some cephalopods have no shell at all.

A skeletal structure in cephalopods that is one of many examples of convergence with vertebrates is the cartilaginous skull. This large rigid structure in the head surrounds and protects the brain. The structure again varies among groups,

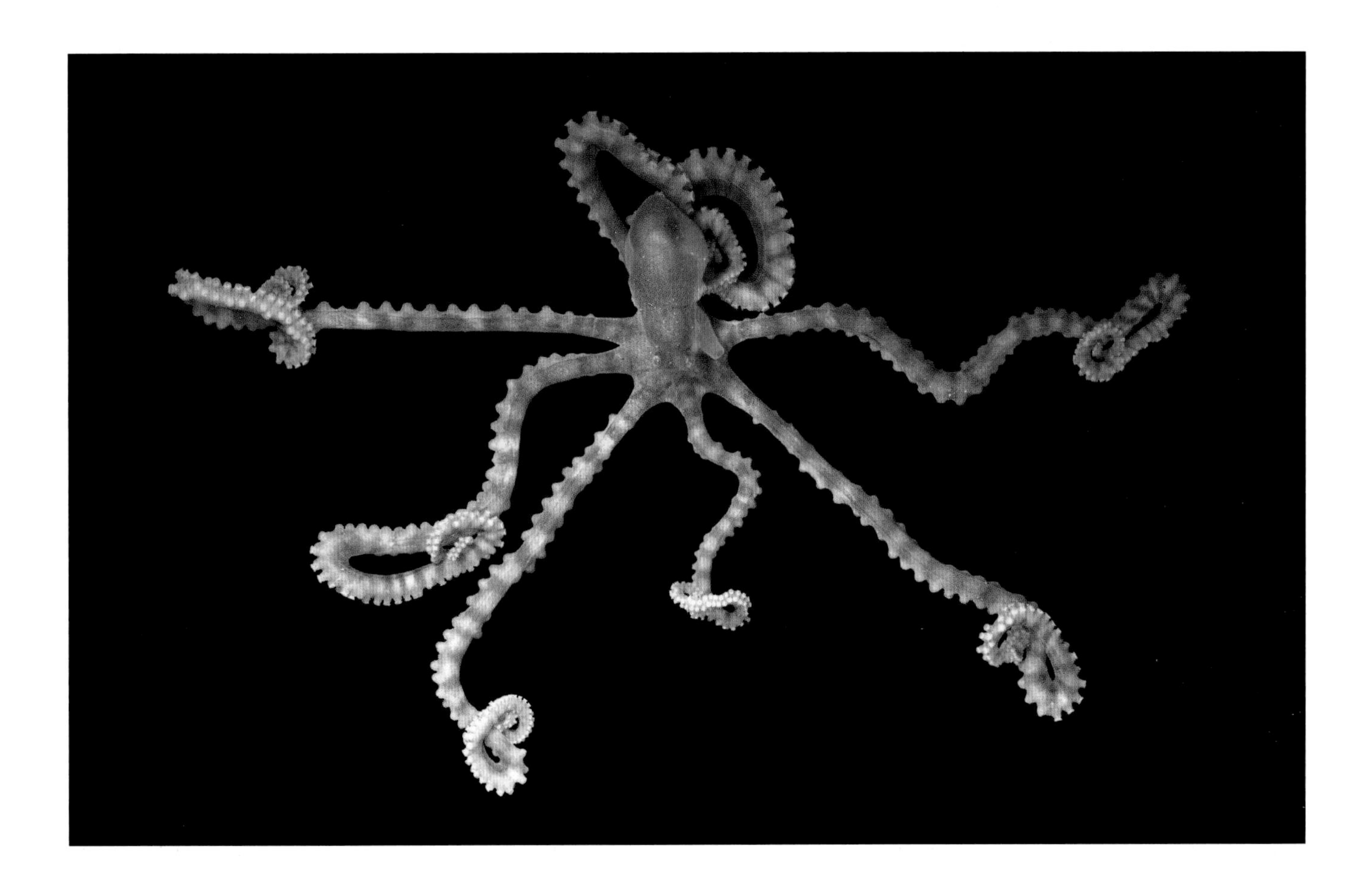

Opposite top X-ray of a nautilid shell showing the septa dividing it into chambers and the perforations of the septa, through which the tissue of the siphuncle controls the amount of water and gas in the chambers.

Opposite middle Ventral (upper) and dorsal (lower) views of a cuttlebone, the internal shell of a cuttlefish.

Opposite bottom In incirrate octopods, the shell is reduced to a pair of stylets or is absent, as in this *Amphioctopus*.

Above The arms of this *Ameloctopus litoralis* move by contraction of opposing sets of muscles without bones, a "muscular hydrostat."

but is composed typically of a dorsal plate, lateral sockets in which the eyes sit, and a posteroventral junction in which the statocysts (see "Other Senses" page 40) are embedded.

A "SKELETON" OF SQUISHY MUSCLE

Cephalopods have no internal bones on which to anchor their muscles, but must nevertheless be able to move their appendages in an almost infinite number of ways. They accomplish this largely by a different mechanical principle known as a muscular hydrostat. A human example of a muscular hydrostat is the tongue. Stick out your tongue and think about how you did it. In cephalopod arms, muscles oriented one way (such as transverse muscles) contract to squeeze other muscles (such as longitudinal muscles) outward in the desired direction. This works because the muscles do not compress, so the total volume remains constant. That is, if the area of a cross-section of the structure is reduced and the volume remains constant, the structure must elongate.

"Hydrostat" refers to squeezing a non-compressible fluid, as an earthworm does to elongate its body. It may be helpful to imagine encircling a toothpaste tube with your hand and squeezing to elongate the contents out through the opening. In the case of the cephalopod arm, the elongated longitudinal muscles can then be contracted to retract the arm, so stretching the relaxed transverse muscles. If the longitudinal muscles on one side of the arm are contracted while those on the opposite side are extended, then the arm is bent in the direction of the contraction. Opposing sets of muscles therefore work on each other rather than on a rigid skeleton to cause necessary movements.

INSIDE THE MANTLE

The mantle is a muscular sac enclosing a seawater-filled space known as the mantle cavity. Within the mantle cavity are the primary organs of circulation, respiration, digestion, excretion, and reproduction.

The respiratory, circulatory, and excretory systems of cephalopods are closely interrelated. When the mantle is opened for a dissection, the internal organs that are most immediately obvious are the lateral, paired, feathery gills. Whereas most living cephalopods have one pair of gills, there are two pairs in the nautilids. Associated with the circulatory system of each gill is a muscular bulb known as a gill heart. These are accessories to the centrally located, three-chambered systemic heart. Thus, when you next get the question in a trivia game, "How many hearts does an octopus have?" the correct answer is three: two gill hearts and a systemic heart.

Below Cross-sections of gills showing how different the structure is among the major groups of cephalopods.

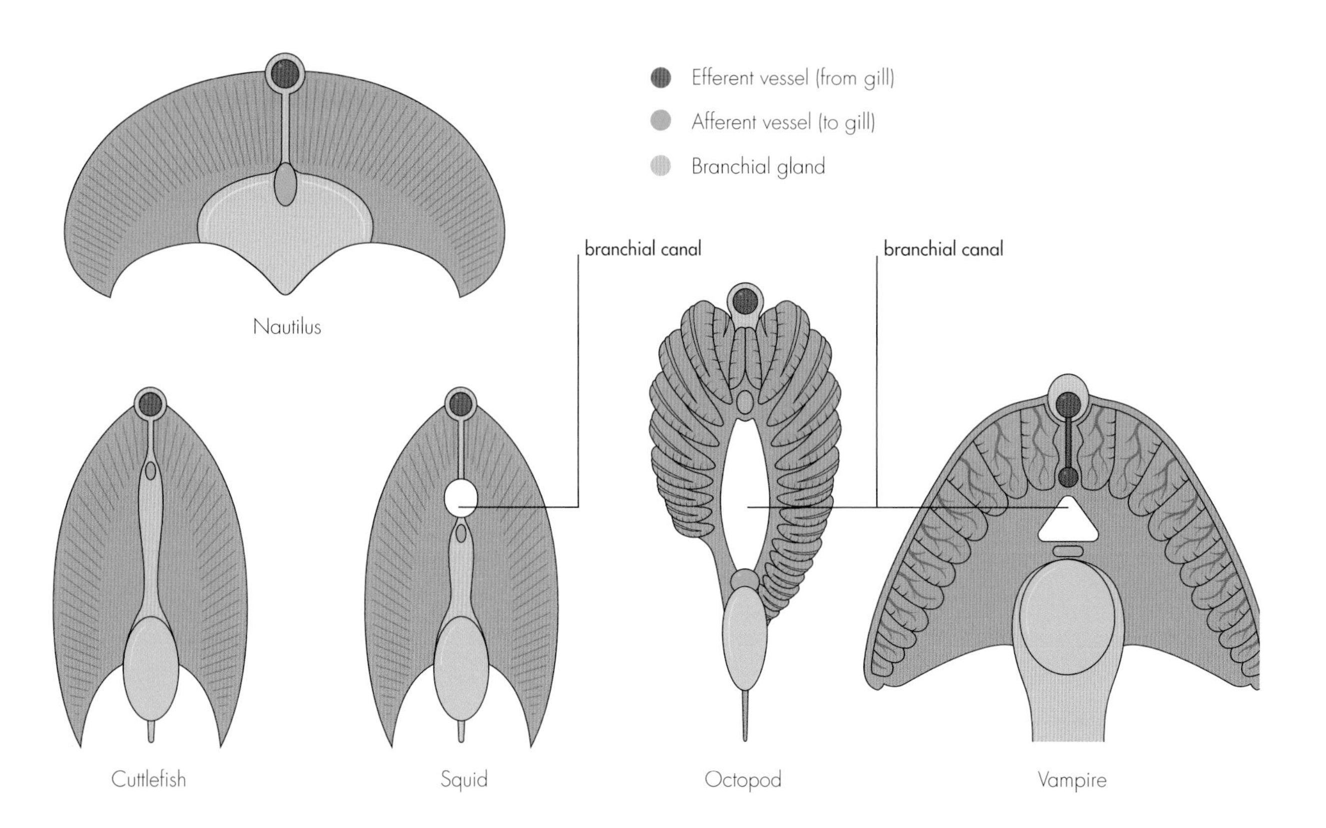

CIRCULATORY SYSTEM

An important characteristic of cephalopod circulatory systems is that the system is "closed." In other words, blood is sent around the body at relatively high pressure through arteries, capillaries, and veins rather than bathing the organs in low-pressure blood flowing through a series of open sinuses, as is found in most molluscs. One characteristic for which the cephalopods have not converged with the vertebrates is the composition of the blood. The chemical that binds oxygen for circulation is copper-based hemocyanin, rather than iron-based hemoglobin, and it is dissolved in the fluid of the blood, not found in special corpuscle cells. Because of the copper base, cephalopods are true blue-bloods; blood with a full load of oxygen is a deep, clear blue in color. This system is not as efficient physiologically as that of red-blooded animals and is likely one of the limiting factors in the competition of cephalopods with fishes.

The oxygenated blood flows from the gills to the systemic heart, which then pumps it out to the various parts of the body through the arteries, with return through the veins. Blood returns from the head region in the cephalic vein, which then splits into two venae cavae leading to the gill hearts. Excretion occurs in outpockets (nephridial appendages) of the venae cavae. These outpockets are surrounded by an epithelium known as the nephridial coelom, which drains into the mantle cavity via a pore, which in some species is elevated on a papilla. The system of nephridial appendages, coelom, and pore is referred to by some as the kidneys. The filtered blood is then pumped by the gill hearts through the set of arteries, capillaries, and veins of the complex folds, sub-folds, and sub-sub-folds that greatly increase the surface area of the gills for exchange of gases with the water in the mantle cavity. After that, the cycle begins again.

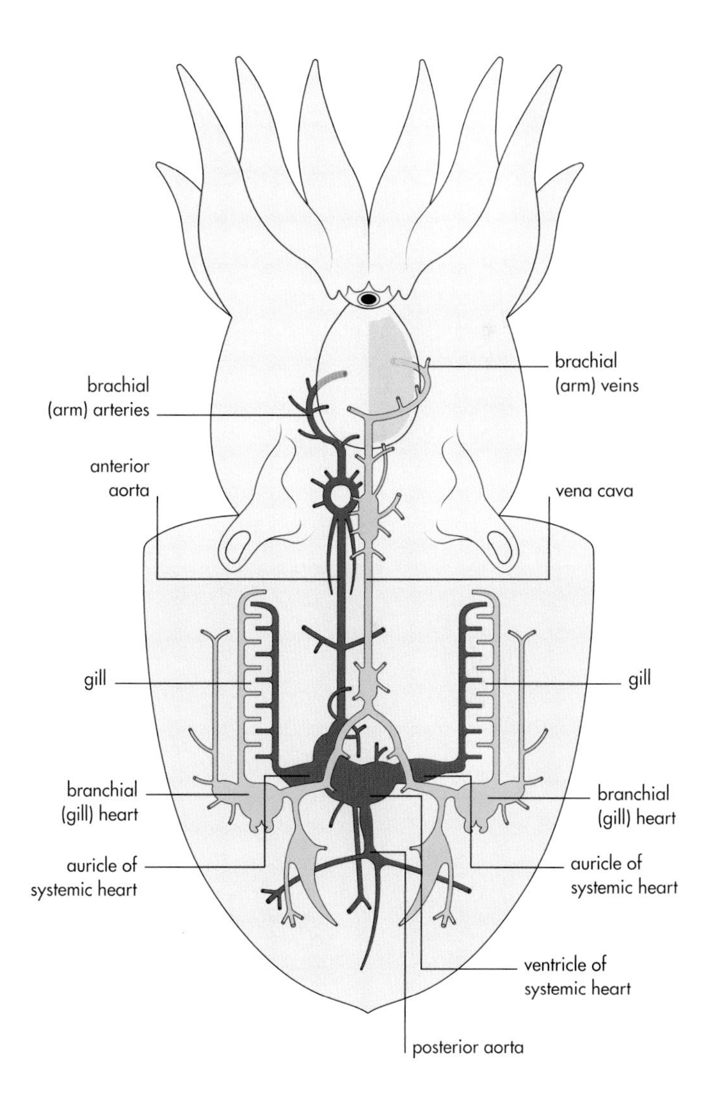

Above The circulatory system of a cuttlefish. Dark blue indicates the systemic heart and arterial system with oxygenated blood. Light blue shows the gill hearts and venous system with deoxygenated blood.

DIGESTIVE SYSTEM

As in most molluscs, the digestive system of cephalopods is U-shaped. The esophagus leads from the buccal mass in the head to its junction with the stomach and cecum in the middle to posterior of the mantle cavity, then the intestine leads back toward the head, emptying into the mantle cavity near the funnel. There are two sets of salivary glands, anterior and posterior, both of which empty via ducts into the buccal mass. The anterior salivary glands are located in the head, usually very close to the buccal mass, and in some cases embedded in it. The posterior salivary glands, also known as poison glands, are usually in the mantle cavity, either closely associated with the digestive gland or embedded in it. Food is reduced by the beaks and radula into small pieces that in some cases are softened by secretions from the salivary glands before the food enters the esophagus. In the anterior esophagus are chitinous plates with backward-pointing barbs that keep the food from moving in the wrong direction. Food is moved along the esophagus by peristaltic muscular contractions, as it is in our digestive system.

Because the brain evolved from the molluscan nerve ring around the esophagus, the food parcels must be small enough to pass through the brain. The esophagus passes through the posterior plate of the cephalic cartilage (skull) and the posterior head into the mantle cavity. There, in some species, it swells for storage of food. This swelling is known as a crop. In some cephalopods, a distinct pouch, called the crop

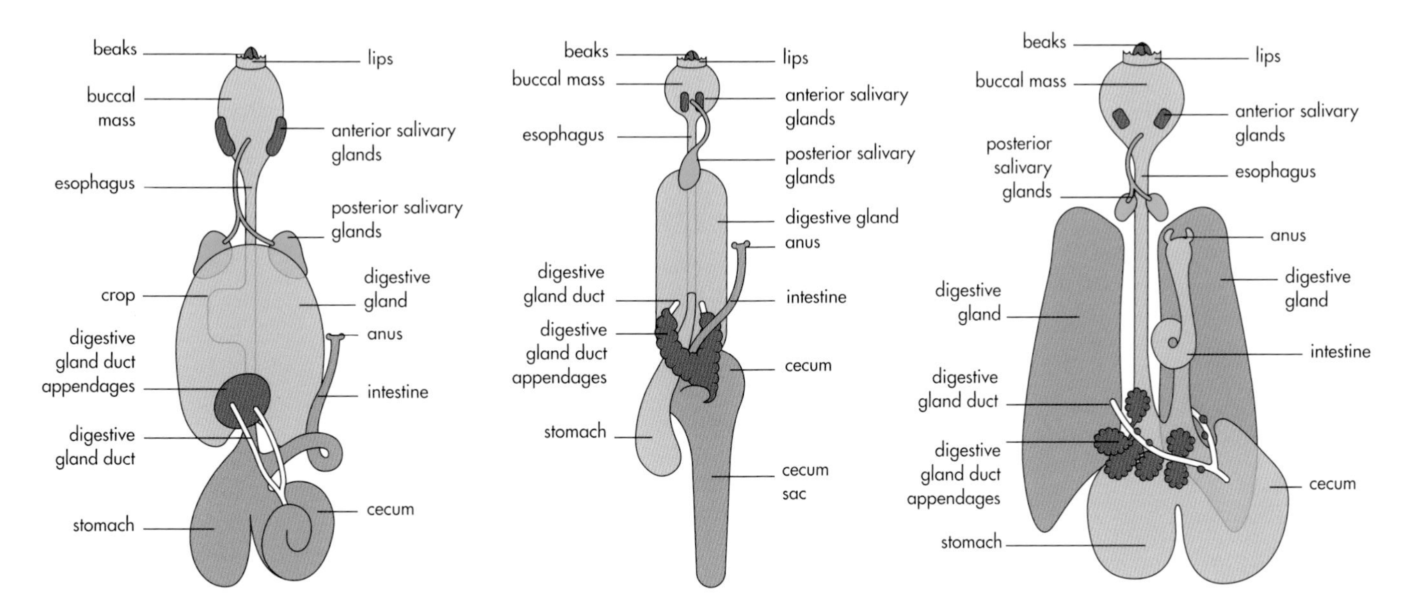

Opposite Because a cephalopod can hold large prey in its arms and bite small pieces with its beaks, the range of prey sizes on which it can feed is generally broad.

Above The shapes and proportions of the organs of the digestive system vary, as seen in these major groups of cephalopods. Left: incirrate octopod, middle: inshore squid, right: cuttlefish.

diverticulum, extends off the crop. Food that has completed transit of the esophagus then enters the single opening of the muscular stomach where primary digestion occurs. The stomach is generally lined with cuticular ridges to aid in grinding food. The stomach may be greatly expandable in size; in species lacking a crop, the stomach serves as the primary organ for food storage. The next stop in the passage of food is the cecum, a primary site of absorption.

Digestive enzymes enter the cecum through ducts from the digestive gland. The cecum tends to have a coiled structure, with ciliated "leaflets" that sort the particles resulting from digestion. This is therefore sometimes called the spiral cecum, to distinguish it from the cecal sac. The latter is the thin-walled posterior portion of the cecum found only in some decapods. Indigestible fragments such as fish bones and pieces of crustacean exoskeleton, are passed directly from the stomach to the intestine. The intestine appears as a simple tube, sometimes with a loop, but has been implicated in water absorption. At the end of the intestine, just inside the posterior part of the funnel, a pair of flap-like anal palps is found in most cephalopods. The function of the anal palps is not well known, but they seem to be involved in the manipulation of ink because the palps are missing in species lacking an ink sac.

Absorption of digestive products also occurs in the digestive gland, which occasionally is referred to incorrectly as the liver. The digestive gland secretes enzymes that pass through the digestive-gland ducts to the cecum. The digestive-gland duct appendages are outpockets of the ducts that are covered with glandular epithelium and are involved in maintaining the ionic fluid balance of the animal. The detailed structure of these appendages differs between octopods and decapods. Fine particles of the digestive products from the cecum pass back up the ducts into the digestive gland, where they are taken into the cells lining the passages of the digestive gland. The final step of digestion by cephalopods is therefore intracellular.

INK SAC

Not involved in digestion, but closely associated with the digestive system, are the ink gland and ink sac. The ink is actually a suspension of melanin produced by the ink gland and stored in the ink sac, which is located between the digestive gland and the intestine. The ink sac empties via a duct into the intestine, from which the ink is expelled through the funnel into the outside world.

In some bobtails and inshore squids the ink sac also has outpockets, which are used to culture symbiotic luminescent bacteria. In addition to the bacterial photophores of the ink sac, intrinsic photophores are found within the mantle cavity on the viscera, including some on the ink sac, in many oceanic squid species.

Below A reef octopus inking. Many cephalopods can eject clouds of ink to help them evade predators.

REPRODUCTION

The sexes are separate in all cephalopods. The unpaired gonads, either ovary or testis, are always located in the posterior region of the mantle cavity. The females of many cephalopods have bilaterally paired oviducts, but in others one of the oviducts is missing (vestigial in a few species). Most males have a single sperm duct on the left side, although a few species have paired ducts. Females have an oviducal gland on each oviduct that secretes a primary coating around the egg; and in many there is a nidamental gland beyond the end of the oviduct, which adds extra protective coatings. Cuttlefishes, bobtails, and inshore squids also have an accessory nidamental gland with a diverse culture of bacteria that is added to the egg tunic to help defend against fungi after the eggs are laid.

In males, the sperm passes through several glands, which package the sperm into a complex spermatophore. The spermatophore is made up of the actual sperm mass, an ejaculatory apparatus that everts the contents of the

spermatophore when it discharges onto the female during mating, and a cement body that attaches the discharged sperm mass to the female. The completed spermatophores are stored in a pouch called Needham's sac, near the penis.

In many, but not all, families of cephalopods, the male has a modified arm, the hectocotylus, which is used to transfer spermatophores to the female. The modifications to the arm take many forms, some quite bizarre. Because hectocotylization occurs very differently and on different characteristic arms in the major taxonomic groups, the presence of a hectocotylus should not be considered a single evolutionary character. For example, the hectocotylus of an octopod is very different from that of a squid. Whether either evolved from the modified arm pair of a nautilid is questionable. Males without a hectocotylus implant their spermatophores with an elongate penis that can extend beyond the mantle edge and funnel. Other forms of sexual dimorphism include features such as greatly elongate arm pairs or tails, and special photophores.

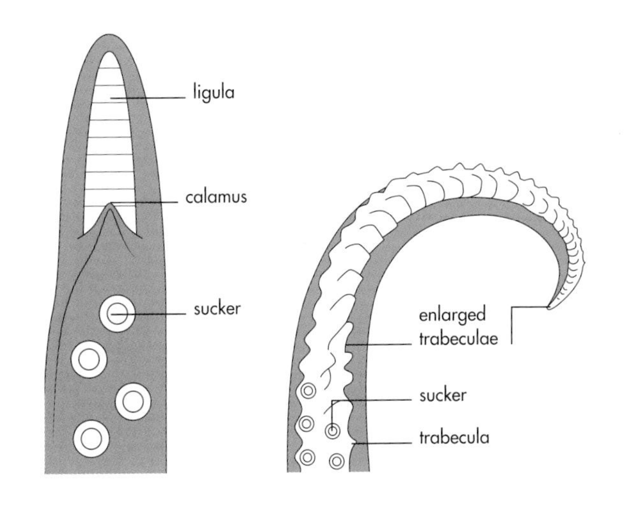

Above Comparison of the modified tips of an octopus (left) and squid (right) hectocotylus.

Below left Mating octopuses.

Below Mating reef squid, *Sepioteuthis*. The male is below, reaching up with the hectocotylus.

Above The advanced nervous system of the octopus enables it to be incredibly agile on a range of surfaces.

A UNIQUE AND COMPLEX NERVOUS SYSTEM

The nervous system and sense organs of cephalopods are remarkable in many ways. The brain is a massive organ of nerve tissue that surrounds the esophagus just posterior to the buccal mass. Many regions of the brain are clearly distinguishable via more than 30 morphologically distinct lobes. *Octopus vulgaris* has half a *billion* nerve cells! By comparison, dogs and cats have only slightly more (about 600,000,000 to 700,000,000).

Yet the brain has only one-third of the total number of neurons in an octopus: two-thirds are distributed throughout the arms, skin, and other organs. This is an extraordinary arrangement, but it is an indication of the importance of the eight arms for locomotion and a wide array of capabilities for sensing and manipulating objects and food. At the base of each arm is a sort of "peripheral mini brain" that can control a few of the arm's functions without central brain control, although overall function and coordination are controlled centrally. The peripheral nervous system also includes the biomedically famous "giant-fiber system" that enables rapid jet propulsion for escape from predators. Some of these fibers (neurons) have enormous diameters: at more than $^1/_{24}$ of an inch (1 mm), the largest in the animal kingdom. Such a bizarre neuronal system for fast locomotion results from an evolutionary dead end. It arose because cephalopods lack Schwann cells, which in vertebrates help to speed up the conduction of nerve impulses. So the only way for cephalopods to get nerve messages to the most distant parts of their mantle as rapidly as they reached the nearer mantle muscles for rapid jetting was to increase the internal diameter of the nerve.

Other highly specialized systems of peripheral nerves (such as arm or brachial nerves, chromatophore nerves, and visceral nerves) control the complex behavior of these remarkable animals. The morphological details and complexity of each of these systems varies among major taxonomic groups.

VISION AND LIGHT SENSING

Most cephalopods have keen visual acuity, can see well in the dark, and can see polarized light, but are color-blind. The similarities in anatomy between the eyes of neocoleoids and those of vertebrates have been mentioned above. Most cephalopod species also have "extra-ocular photoreceptors" that are capable of detecting light. In decapods they are found within the cephalic cartilage whereas in the octopods they are near the stellate ganglia inside the mantle. These organs are sensitive to bioluminescent and other down-welling light. In some squids, the photoreceptors are located near photophores, enabling the squid to monitor the amount of light that it is producing in addition to the amount of down-welling light from the sun.

Recently a novel discovery found opsin molecules—the same molecules found in the retina that detect light—throughout the skin of cuttlefishes, squids, and octopuses. Their function is not known but the likely explanation is that the skin can somehow detect light and that this can perhaps aid some aspect of camouflage, such as matching brightness to the surrounding substrate.

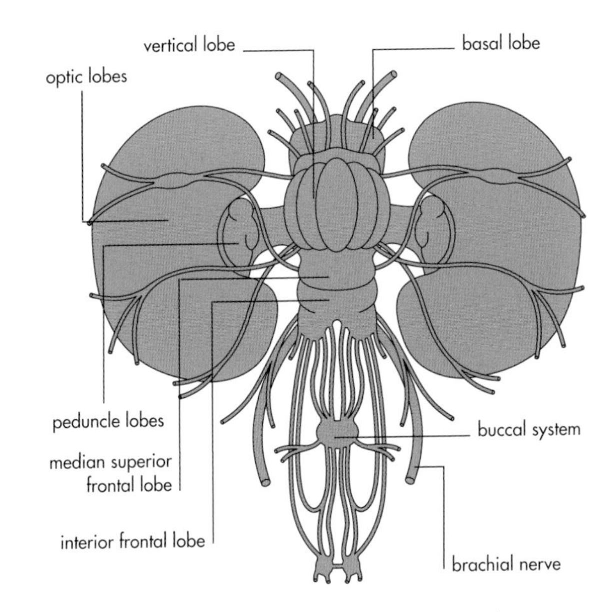

Above Diagram of the octopus brain showing the complexity of the multiple lobes and innervation.

Below left The arms of this benthic octopod have more neurons than are in its brain and are capable of independent movements.

Below This opalescent inshore squid, *Doryteuthis opalescens,* is using its chromatophores to match the brightness of the sand around it.

OTHER SENSES

Cephalopods can monitor gravity as well as changes in their orientation and motion through a sensory system composed of a pair of statocysts, which contain a tiny calcium-carbonate stone attached by hair-like sensory cilia to a pocket in the statocyst. Inertia of the statolith not only informs the animal which way is up but also detects linear acceleration ("Am I speeding up or slowing down?"). Other sensory cilia project into the fluid within the statocyst and detect rotational changes through relative movements of the internal fluid.

Depending on the family and species, various knobs and ridges project into the fluid, constraining the directions in which it can move. The result is a system that is functionally similar to the circular-canal system in vertebrates, a very sophisticated system for detecting angular acceleration, which must be important for the fast, non-uniform motion of cephalopod jet propulsion.

Above To maintain control when swimming, this Caribbean reef squid (*Sepioteuthis sepioidea*) must be able simultaneously to sense gravity, acceleration or deceleration, and rotation in various directions.

Recently, a novel sensory system has been discovered in decapods, a mechanoreceptor system that is functionally similar to the lateral-line system in fishes. These lines of epidermal hair cells are sensitive to water movement and can detect a fish swimming tens of yards away.

Hearing has also been determined recently in cephalopods, but unlike terrestrial animals that sense pressure waves of sound, they sense the particle motion aspect of aquatic sound. This sensitivity is in the range of "low frequency" vibrations (ranging from 1 to 500 Hz) but they are insensitive to the ultrasonic clicks of echolocating whales and dolphins, and so are "deaf" to those approaching predators. The statocysts provide this hearing capability.

Above left A mating pair of the opalescent inshore squid *Doryteuthis opalescens* with egg cases.

Above Benthic octopods have chemosensory cells on their suckers and seem able to taste by touching objects with the suckers.

Taste is the ability to detect, categorize, and react to chemicals when an object is touched. Chemosensory cells on the surfaces of suckers, and behavior consistent with this explanation, have been demonstrated clearly for octopods. This is probably also true for decapods. Touch is accomplished by suckers in cephalopods; these have mechanoreceptors that can discriminate physical texture. Recent research has discovered a surprising sensory response: when male squids touch egg cases they detect a contact pheromone that elicits strong aggression between males as they compete for female mates on spawning grounds.

Smell ("olfaction" or odor detection) for a marine organism is a skill equivalent to taste for chemicals that are dissolved in seawater and so is a more long-distance sensing system. Sensing chemicals is very important in a fluid environment, especially for animals that are nocturnal or live in the deep sea. A bilateral pair of organs on the head has long been suspected of chemosensory function, and is called the olfactory papillae. The position varies among groups, from containment in pits to elevation on stalks. Fairly recently, the olfactory function of these organs has been confirmed experimentally. Particularly important may be the detection of ink from members of the same species as a signal of alarm.

Finally, there is the possibility of a sensory system that has not yet been found. Some ommastrephid squids are known to migrate long distances. In vertebrates with similar migrations a magnetoreceptor system is used for navigation based on the earth's magnetic field. The presence of a similar system has been postulated for migratory cephalopods but has yet to be discovered.

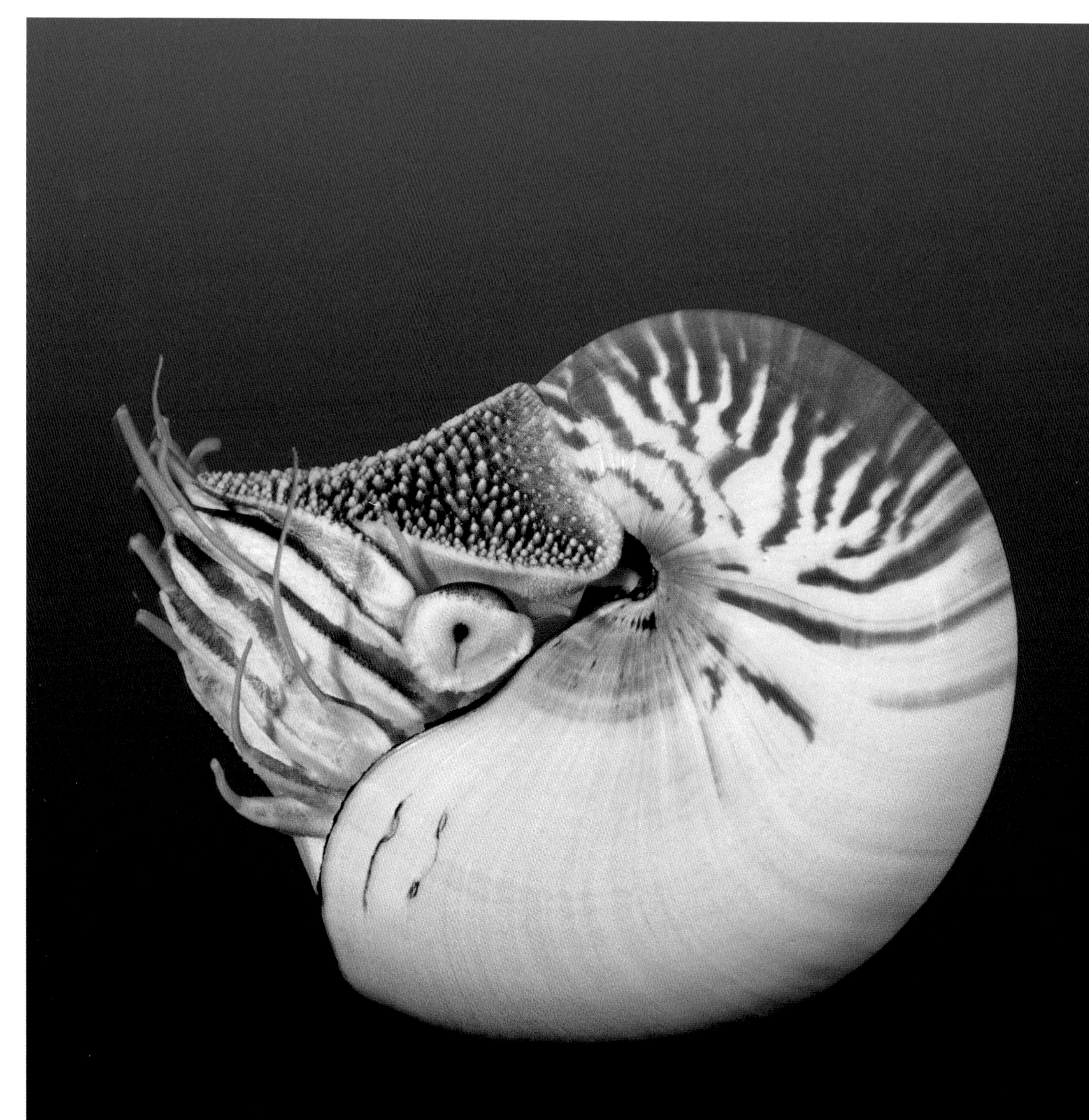

EMPEROR NAUTILUS

Nautilus pompilius

FAMILY	Nautilidae
OTHER NAMES	Common nautilus, flame nautilus
TYPICAL HABITAT	Continental shelf and slope in the "coral triangle" area of the Indo-West Pacific. Depth to about 2,460 ft (750 m) depth
SIZE	Shell diameter up to at least 9½ in (24 cm)
FEEDING HABITS	Scavenging and opportunistic predation
KEY BEHAVIORS	Vertical migration close to deep reef slope, limited by ability to compensate for effects of pressure changes on gas-filled chambers

ONE OF THE FIRST CEPHALOPODS DESCRIBED by modern taxonomy, this species has long been considered a classic example of a "living fossil." *Nautilus pompilius* is the most widespread of the living nautilid species, but large variability among local populations and seemingly low gene flow have resulted in proposals that this wide distribution comprises a complex of closely related species. Some of the populations have been strongly affected by commercial harvesting for the shell trade and for meat. The scavenging habit of nautilids makes them easy to catch in baited traps. In 2016, all living nautilid species were granted international protection under the Convention for International Trade in Endangered Species (CITES).

NOT NECESSARILY A LIVING FOSSIL

Very little fossil evidence of soft-tissue anatomy exists for the ancestral nautiloids. So, although the nautilus shell is similar to abundant nautiloid fossils, the soft anatomy may be different. In living animals, in addition to the external shell, the "pinhole camera" eyes and many arms easily distinguish nautilids from other cephalopods. Not visible externally, the two pairs of gills are also quite different from those of other cephalopods. Fossil nautiloids were long considered to be "tetrabranchiate" (having four gills) but this could be a fairly recent duplication in the evolution of recent nautilids.

A CHAMBERED SHELL

Although the large external shell of a nautilus is reminiscent of a snail shell, most of the nautilus shell contains a series of chambers, limiting the area into which the body can be withdrawn. When the cephalopod pulls its head and body into the shell, water is expelled, providing a jet for propulsion.

GIANT PACIFIC OCTOPUS

Enteroctopus dofleini

THE GIANT PACIFIC OCTOPOD, SOMETIMES called simply GPO, is among the largest octopods. It is also among the best known cephalopods. Because the species lives well in captivity, it is popular for display in public aquaria. They are easily found by divers in the North Pacific and have been the subject of television documentaries. Therefore, many behavioral observations have been reported for this species. Additionally, GPOs are harvested commercially both for seafood and as bait for other fisheries. Recent genetic studies have shown the giant Pacific octopod and some deep-sea relatives to be so different from typical shallow-living species of octopus that they belong in distinct families.

HOW GIANT ARE THEY?

How large a cephalopod species gets in the wild is difficult to determine. Although scientific literature documents that this species is indeed a giant, anecdotal reports sometimes mention even larger sizes. Undocumented reports mention octopods with arms spans exceeding 20 ft (6 m).

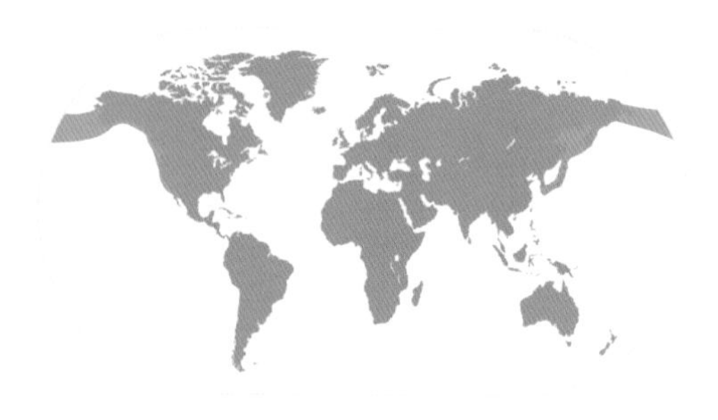

FAMILY	Enteroctopodidae
OTHER NAMES	North Pacific giant octopus
TYPICAL HABITAT	Lives on the bottom from the intertidal to depths greater than 4,920 ft (1,500 m), rock and sandy shell
SIZE	Mantle length to 24 in (60 cm), total length to nearly 10 ft (3 m)
FEEDING HABITS	Typically eats molluscs and crustaceans but sometimes catches fishes and even birds
KEY BEHAVIORS	Very flexible behavior, both in the wild and in captivity

VARIED HABITATS

Although often found in shallow water around the North Pacific, the species is also found in very deep water. Its ability to texture its skin with flaps and ridges allows it to blend into diverse habitats.

PHARAOH CUTTLEFISH

Sepia pharaonis

FAMILY	Sepiidae
OTHER NAMES	Seiche pharaon
TYPICAL HABITAT	Stays on or near the bottom at depths to about 430 ft (130 m)
SIZE	Mantle length to 16½ in (42 cm)
FEEDING HABITS	Feeds at night on crustaceans and small fishes
KEY BEHAVIORS	Seasonal migration to form reproductive aggregations

COMMON IN WARM SHALLOW WATERS throughout the western Pacific and Indian Oceans, this species is important to fisheries, rearing, and experimental studies. It has been raised through the entire life cycle in captivity. As with many cephalopod species with apparently broad geographical distributions, the pharaoh cuttlefish is actually a complex of closely related and morphologically similar species. Their abilities to change patterns of color and texture rival those of shallow-water octopods.

COMPETITIVE MALES

Males fight for preferred mating locations and for access to females in the areas where the species congregates for reproduction. Such fighting may include behavioral displays of color pattern and texture, or threatening actions such as shoving of rivals.

FINS FOR EASY MOVEMENT

Cuttlefishes, such as the pharaoh cuttlefish, can easily hover above the bottom because the gas-filled chambers of the cuttlebone provide neutral bouyancy. Because the fins extending along the sides of the mantle are capable of variable independent patterns of undulation and flapping, cuttlefishes are very maneuverable.

BIGFIN SQUID

Magnapinna pacifica

ALTHOUGH ONLY RECENTLY DISCOVERED, BIGFIN squids have been seen repeatedly from both manned and robot submersibles diving to depths of at least 3,300 ft (1,000 m) in many parts of the world's oceans. Because of their very long arms and tentacles, they are among the largest deep-sea nekton. Despite their charismatic appearance and repeated observations, almost nothing is known about their biology. The long arm tips have microscopic suckers and are held dangling downward from the elbow-like junction with the main arm. They are assumed to stick to any zooplankton that bumps into them, but how such prey is then transferred to the mouth is unknown. The species was named *Magnapinna pacifica* because the first few specimens were caught in the Pacific Ocean. However, we now know that it lives in the deep Atlantic as well, so the specific name is not as appropriate as was first thought.

NEW FAMILY RECOGNIZED FROM UNUSUAL PARALARVAE

Although the long spaghetti-like tips to the arms and tentacles are the most obvious feature of magnapinnids, they are not the reason the family was initially recognized. The family was described from two small juveniles and a paralarva, none of which had yet developed the long arm tips. These small squids did not match the characteristics of any known family. Because they all had very large fins, though, the squids were named magna pinna, "big fin".

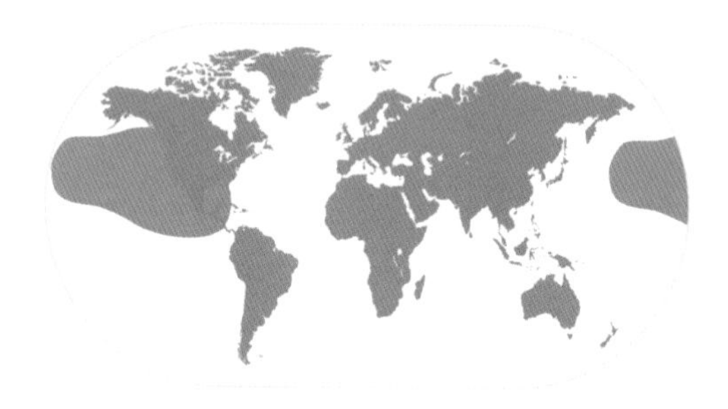

FAMILY	Magnapinnidae
OTHER NAMES	Long-arm squid
TYPICAL HABITAT	Deep-sea, bathypelagic
SIZE	Mantle length to about 12 in (30 cm), total length to about 23 ft (7 m)
FEEDING HABITS	Presumably catches small crustaceans or fishes with microscopic suckers on arm-tip extensions
KEY BEHAVIORS	Drifts with arm-tip extensions dangling downward from elbow-like junctions with main arms and tentacles

LONG ARM TIPS

Bigfin squids swim not by jetting, but by flapping the fins from which their common name derives. As the squid flaps away, the extremely long arm-tip extensions drag behind; the tips can be retracted to much shorter lengths than when they are deployed for feeding. However, if an arm tip contacts something, such as a submersible, it sticks and the animal seems to have difficulty letting go.

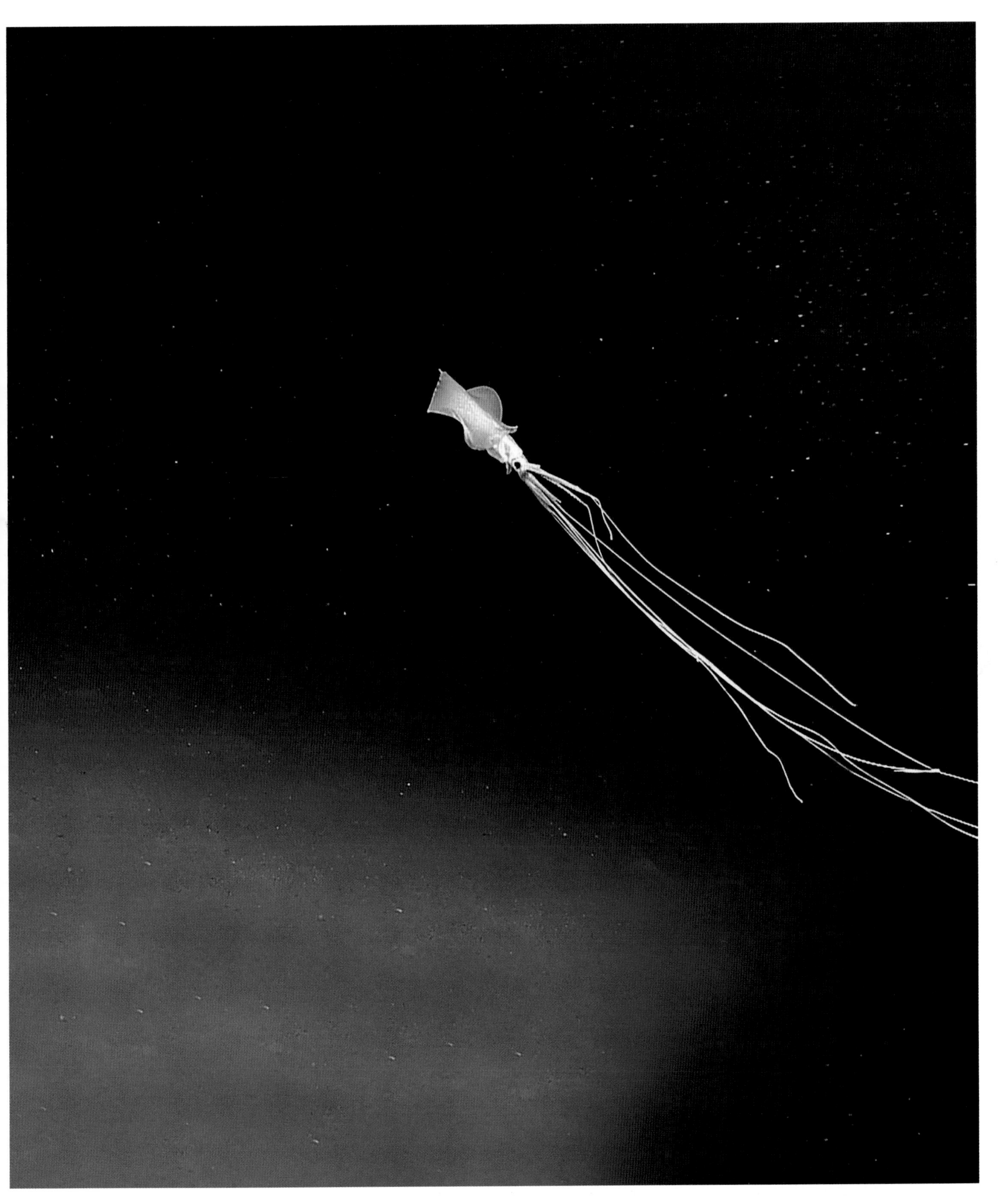

JAPANESE FLYING SQUID

Todarodes pacificus

FAMILY	Ommastrephidae
OTHER NAMES	Surume-ika, among other local names
TYPICAL HABITAT	Continental shelf and slope in the North Pacific
SIZE	Commonly to 12 in (30 cm) mantle length, reported to 20 in (50 cm)
FEEDING HABITS	Shifts with growth, from copepods to shrimps to fishes
KEY BEHAVIORS	Juveniles can jump clear of the water and glide similarly to a flying fish

THE JAPANESE FLYING SQUID IS THE TARGET of one of the largest cephalopod fisheries worldwide. Because of its importance as a commercial fishery resource, this species is also one of the best-studied cephalopods, especially among the oegopsid squids and other oceanic cephalopods. Actively swimming throughout the water column at night over the continental shelf and slope, they descend to the bottom during the day. Schools of them have been studied in large seawater tanks. This has allowed observations on reproduction that could not be made in the wild. As do other ommastrephids, spawning *Todarodes pacificus* females produce multiple egg masses. These are large (3 ft (1 m) or more in diameter) gelatinous spheres with eggs dispersed throughout the jelly. The egg mass is released into the open water column where it settles to a depth at which it is neutrally buoyant. There it drifts for the few days necessary for the embryos to develop to the very small (mantle length less than $^{1}/_{24}$ in (1 mm)) hatching stage.

LONG-DISTANCE MIGRATION

The life history is closely associated with the western boundary current system (Kuroshio and Oyashio) east of Japan and with the Tsushima warm current in the Sea of Japan. The species undertakes a long-distance migration southward to spawn. Egg masses and paralarvae drift northward as they develop. Major feeding grounds are in the northern part of the range.

THREE POPULATIONS

Although the life span of the Japanese flying squid is only about a year, three distinct populations, separated by spawning season, are recognized. In spite of some differences in morphology among populations, gene flow among them has been confirmed.

ROUGH GLASS SQUID

Cranchia scabra

FAMILY	Cranchiidae
OTHER NAMES	Rough cranch squid
TYPICAL HABITAT	Epipelagic when young, meso or bathypelagic subsequently
SIZE	Mantle length to 6 in (15 cm)
FEEDING HABITS	Not known
KEY BEHAVIORS	Bioluminescence; inking in mantle

THE ROUGH GLASS SQUID IS A SMALL SQUID found worldwide in tropical and subtropical waters. It can produce bioluminescence and has a series of photophores around the eyes. Mature females also have photophores on the tips of their arms, which are probably used to attract mates. The mantle of *Cranchia scabra* is covered in small cartilaginous tubercles—hence "rough" glass squid—and they are able to withdraw their head and arms into the mantle cavity, turning themselves into a knobbly spherical ball—presumably as a defensive mechanism. They may also ink into their own mantle cavity when threatened, giving themselves a more opaque appearance. This defensive behavior is not well understood but may provide them with some camouflage against their predators. However, rough glass squids can use their chromatophores to go completely and instantaneously dark, so maybe the ink, which leaks slowly from the mantle, acts as a chemical defense or deterrent. Their predators include large species such as short-finned pilot whales and sperm whales, and sea birds such as albatrosses and petrels.

CHEMICAL BUOYANCY

All glass squids have an enlarged body chamber, the coelom, filled with ammonium chloride. This solution is less dense than the squid's other tissues, and provides the squid with overall neutral buoyancy, such that it floats effortlessly. To pass water over their gills, rather than contracting their mantle, they pass peristaltic waves through the coelom.

TRANSPARENCY FOR CAMOUFLAGE

Cranchia scabra spends its early life in shallower waters that are infused with light. The transparent body, characteristic of glass squid species, helps provide camouflage in this environment. When they are around 4¾ in (12 cm) long, after about four months in epipelagic waters, they change their lifestyle and inhabit deeper waters, possibly descending as far as 6,600 ft (2,000 m) depth.

PHYLOGENY & EVOLUTION

500 MILLION YEARS OF EVOLUTION

ALTHOUGH ANIMALS CLEARLY RECOGNIZABLE as modern octopods, squids, and cuttlefishes are not found in the fossil record until at least the Cretaceous period, cephalopods have their origins much earlier in Cambrian Seas. Ancient cephalopods had external shells, a feature that in recent cephalopods is only found in nautiluses. These calcified shells preserved well on the ancient seafloors and the fossil record of cephalopods from the Paleozoic and Mesozoic Eras is arguably richer than that from less ancient times.

A SHELLED ANCESTRY

There are eight classes of living mollusc. Some are well known, such as cephalopods, gastropods (snails), and bivalves (clams, mussels), and some are less so, such as Monoplacophora. Externally resembling a limpet, with a soft body and a hat-like shell, but with serially repeated internal organs such as gills and nephridia (primitive kidneys), Monoplacophora are thought to be the closest living relatives of cephalopods. Early cephalopods probably evolved from a monoplacophoran-like ancestor in the Cambrian period.

THE EARLIEST CEPHALOPODS

There were early cephalopod ancestors living on the seafloor of the shallow Cambrian seas and true cephalopods by the late Cambrian—more than 490 million years ago. *Plectronoceras cambria* is the earliest undisputed true cephalopod. Although these creatures were very small—less than ½ in (1.25 cm) tall—they had the same external chambered shells as their better-known cousins so abundant 50 to 100 million years later. *Nectocaris*, which was found in the famous Burgess Shale of the Canadian Rocky Mountains and dates from 508 million years ago, has been proposed as an even earlier cephalopod. But this has been strongly disputed. *Nectocaris* lacks features common to all molluscs, such as the file-like tongue and the radula, and common to all cephalopods, such as beaks, and probably belongs to one of the many animal phyla that evolved in the Cambrian but went extinct long ago.

Top Ammonite with sutures.

Middle *Lituites* were shelled cephalopods from the middle Ordovician with a planispirally coiled shell apex, expanding into a long, almost straight shell.

Bottom The nautiloid *Orthoceras* had a straight but gently expanding, elongate shell.

THE GREAT ORDOVICIAN BIODIVERSITY EVENT

Although the Cambrian Explosion is famed for the origin of new, spectacular, sometimes bizarre body plans, and the origin of most of the animal phyla, the Great Ordovician Biodiversity Event involved the greatest overall increase in biodiversity in geological history because of an increase in diversity within phyla. The radiation of cephalopod fauna at this time was no exception. Cephalopod diversity was limited at the Cambrian–Ordovician border. These cephalopods were relatively small, with slightly curved shells, and were restricted to shallow seas, probably living on or close to the seafloor. By the end of the Ordovician, just 50 million years later, cephalopods had expanded their range to higher latitudes, and developed shells with greater buoyancy and resistance to implosion, allowing them to inhabit the water column and live farther from the shore.

RISE OF THE AMMONITES

The early Devonian period, 420 million years ago, heralded the arrival of both the nautilids (ancestors of living nautiluses) and perhaps the most famous group of fossil cephalopods, the ammonites. Both groups were hugely successful. Ammonites share many morphological features with nautiluses, including the chambered shell, but the detailed and highly ornamented sutures between the shell chambers are characteristic of ammonites and make them easy to distinguish as fossils. Ammonites diversified widely in shallow waters, and persisted for more than 350 million years to the end of the Cretaceous. However, individual species of ammonite endured for only a few million years and they can be used as "index fossils" to date rock strata. Like modern octopods, squids, and cuttlefishes, ammonites were predators. They probably hunted in competition with the newly jawed fishes that arose at the same time. Ammonites themselves arose from straight-shelled Devonian cephalopods called bactritids. The bactritids did not last much beyond the Permian, but their legacy lives on, as they are also the likely ancestors of modern shell-less coleoids—the octopods, squids, and cuttlefishes.

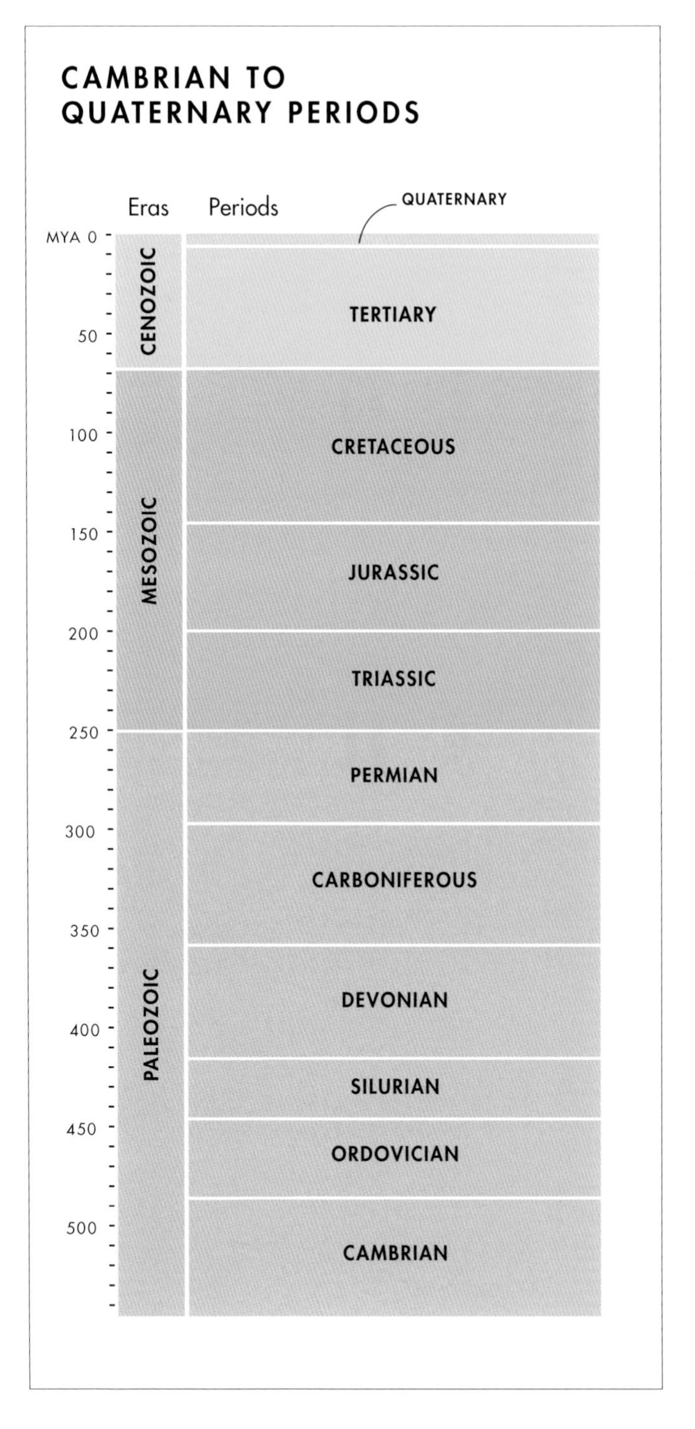

NAUTILUS—LIVING FOSSIL?

LIVING NAUTILUSES COMPRISE JUST A handful of species in two genera, *Nautilus* and *Allonautilus*, which are found on the slopes of coral reefs in the Indo-Pacific. Their ancestors can be traced back 400 million years through periods when their diversity and numbers were much greater. The shells of modern nautiluses are similar to those from the Jurassic seas 200 million years ago, but their internal anatomy may have evolved in that time.

LIVING NAUTILUSES

Nautiluses are unlike any other living cephalopod. They retain the external, chambered shell of ancient cephalopods and have simple "pinhole-camera" eyes, which are less sophisticated than the eyes of coleoid cephalopods. Nautiluses hatch as miniature adults, with shells about 1¼ in (3 cm) diameter,

Below *Nautilus pompilius*, the chambered nautilus, seen here in the waters of the Great Barrier Reef, Australia.

from eggs attached to rocks by females. They secrete their shell as they grow, always living in the most recently formed chamber, and sealing subsequent chambers with dividing walls called septae. The constant addition of gas-filled chambers allows the animal to control its buoyancy and compensate for the heavier weight of adults, which grow to around 8 in (20 cm) in diameter.

Below *Allonautilus scrobiculatus*, the fuzzy nautilus, inhabits the waters around New Guinea.

ORIGINS

It is uncertain from which branch of early cephalopods the nautilids evolved, but they first appeared in the early Devonian, about the same time as the ammonites. It was perhaps at this time that nautilids showed their greatest diversity of form with species that were tightly coiled or "planispiral" like modern nautiluses, species that were only loosely coiled, or species that were simply curved like some of the earliest shelled cephalopods. They reached their greatest

species richness in the Carboniferous period, 360–300 million years ago. By this time, nearly all their shells were planispiral, likely because these were easier to maneuver, particularly in larger animals. Nautilids survived all four major mass extinction events, including the one at the end of the Cretaceous that wiped out the ammonites. Although nautilids suffered declines in diversity at each extinction event, they always recovered and re-diversified. Just a single genus, *Cenoceras*, survived the Triassic–Jurassic mass extinction 200 million years ago, and yet nautilids subsequently diversified to give rise to some of the largest ever known—reaching diameters greater than 20 in (about 50 cm). Once again, this diversity was lost and Jurassic genera went extinct, leaving a single lineage that gave rise to the few species known today. The shell of *Cenoceras* is very similar to that of modern species, suggesting that externally nautiluses have changed little in 200 million years. This extraordinary consistency in shell form has sometimes led to nautiluses being described as living fossils. However, biologists argue that their internal anatomy has been less consistent and this may be a misnomer.

BIOLOGY

Nautiluses are much longer lived than most cephalopods. It takes a year for their eggs to hatch, another 15 years for juveniles to reach maturity, and they survive at least 5 years after maturing. Their whole life is led at a slower pace. They inhabit the fore-reef slopes of coral reefs migrating from 230 to 2,300 ft (70 to 700 m), below which depth their shells implode.

Opposite left Cutaway of a nautilus shell, showing the large living chamber, and numerous smaller buoyancy chambers.

Opposite right A fossil of *Cenoceras*, the only nautilus genus to survive the Triassic–Jurassic extinction.

Right Shell of *Allonautilus* demonstrating its morphological similarity to the Jurassic *Cenoceras*.

Their migrations are probably controlled by a combination of needs: to forage, to retreat from predators, and to rebalance their buoyancy chambers, which suffer a slow influx of fluid when the external pressure becomes too high.

Although nautiluses move by "jet propulsion" they are not fast. The term reflects not speed, but a mechanism by which nautiluses push water through their hyponome—two muscular flaps that unite to perform the same function as the funnel of a squid or octopus. By varying the orientation of the hyponome they are able to move in different directions through the water column. Close inspection of nautilus shells reveals scars where the hyponome muscles attach. Ancient nautilid and ammonite shells bear similar muscle scars, suggesting that ancient shelled cephalopods were also jet propelled and may have had similar habits.

POPULATION DECLINE

Trapped in baited pots for their beautiful shells and meat, nautilus populations in some areas have declined precipitously and conservation efforts are required in some parts of the Indo-Pacific. More than 1.7 million nautilus shells were imported into the USA alone in the 15-year period ending in 2015. In 2017, the Convention on International Trade in Endangered Species (CITES) listed Nautilus in Appendix II, which includes species that could face extinction if trade is not controlled. Nautilus shells can no longer be exported without a permit.

EVOLUTION OF COLEOIDS

COLEOIDEA—CEPHALOPODS WITH INTERNALIZED shells—arose in the Carboniferous or Permian. Their ancestors are the straight-shelled bactritids that also gave rise to the ammonites. Coleoidea includes not only all living octopods, squids, and cuttlefishes, but also extinct forms such as belemnites. The eight-armed Octopodiformes that include modern octopods arose in the Triassic. Decapodiformes, which have ten arms, two of which are modified into tentacles, and include squids and cuttlefishes, arose in the Jurassic.

BELEMNITES

All living cephalopods found in the oceans today, with the exception of the nautiluses, are coleoids. Hence all octopods, squids, and cuttlefishes are part of this group that arose more than 250 million years ago. Not surprisingly, there are some more ancient members of this group that have not survived to the present. The group with the richest fossil record is probably the belemnites, which roamed the seas in the Jurassic and Cretaceous. As in all coleoids, the shells of belemnites were internal, but they were less reduced than the shells found in modern-day squids. The anterior part of the shell was reduced and supported the mantle, the middle part was the chambered phragmocone that provided buoyancy, while the posterior "guard" of the shell was well developed and probably provided a counterbalance to the heavier head end of the animal. Fossil belemnite guards are prevalent in many strata, and other shell parts are not infrequent.

In contrast, fossils where the soft tissues have also been preserved are relatively rare, but the examples we have, almost all of them from Lagerstätten—rich fossil beds of exceptional preservation—reveal information about arms and internal organs.

Below left *Ostenoteuthis* is a coleoid cephalopod from the early Jurassic.

Below By measuring oxygen isotope ratios in belemnite fossil rostra, geologists can estimate the temperature of the seas in which they lived.

Opposite Diagram of belemnite showing the proostracum supporting the mantle, the chambered phragmocone, and the heavy rostrum.

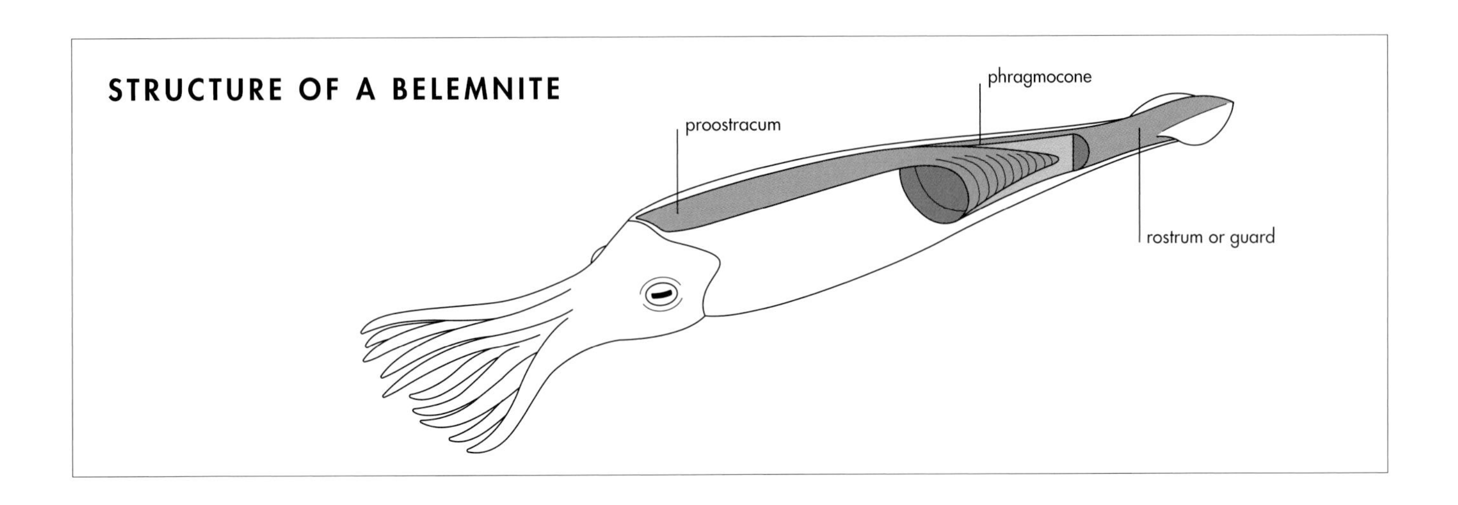

COLEOID ADVANCES

Cephalopod-yielding Lagerstätten include the Carboniferous Mazon formation in Illinois, USA, which has yielded some of the earliest coleoids. The Oxford Clay in southern England and Solnhofen limestone beds in southern Germany—the latter famous for the discovery of *Archaeopteryx*—have together yielded huge Jurassic diversity. Exceptionally preserved fossils show that these early coleoids had many advanced features. They all display the streamlined body that we associate with recent squids, but some fossils even show that their mantle muscle was developed enough to make them powerful swimmers in contrast to the ammonoids and nautiloids with which they co-existed. They had arms with hooks, although these structures probably did not have the same developmental origins as the hooks of modern species of squid, and they had beaks to tear at their prey.

Ink sacs are discernible in many Jurassic cephalopod fossils, so we know that, just like their modern counterparts, early coleoids used inking as part of their defensive systems. Scanning electron microscopy has even revealed this Jurassic coleoid ink to be chemically similar to modern cephalopod ink. It is not known whether early coleoids had chromatophores—muscle-encircled pigment sacs used to generate the spectacular visual displays on the skin of modern cephalopods—but certainly these evolved before the divergence of the Decapodiformes in the late Jurassic. Unquestionably it was the internalization of the coleoid shell that exposed large amounts of mantle tissue for the first time. Early coleoids also had eyes with lenses, a feature found in octopods, cuttlefishes, and squids, but absent from nautiloids and probably ammonoids. This improved vision was almost certainly essential in prey capture and predator avoidance and may have helped in the development of signaling.

These advanced features probably all contributed to coleoid success. Their reduced shell, hydrodynamic shape, and stronger mantle made coleoids faster, and their improved sensory and motor systems improved their competitive edge against other highly successful groups such as fishes. Doubtless these characteristics helped coleoids to survive three mass extinctions, including the last great mass extinction at the end of the Cretaceous period, which ended not only the dinosaurs, but also the ammonoids, the other great cephalopod lineage.

THE BUOYANCY CONUNDRUM

UNTIL AT LEAST THE CARBONIFEROUS, ALL cephalopods were hindered by an unwieldy external shell. Perversely, the very shell that had provided buoyancy and allowed Ordovician cephalopods to first conquer the water column became an impediment. It limited speed and the maximum depth that could be occupied. The internalization of the shell in coleoids was the first step to a solution, but over the next 250 million years, cephalopods evolved multiple solutions to establish themselves in the world's oceans.

ANCIENT SHELLS

Ancient cephalopods had large cumbersome external shells, in keeping with their molluscan ancestry. These shells, made of calcium carbonate, were distinguished by a series of gas-filled chambers termed a phragmocone, still seen in living nautiluses (see species profile p86). The animals lived in the largest and most recently formed chamber, and the outgrown smaller chambers acted as reservoirs for gas and fluid. A narrow tube of tissue, the siphuncle, penetrated the chambers, extracting fluid and replacing it with gas. The siphuncle was positioned differently in ancient cephalopods such as ammonites, but it almost certainly functioned as it does in nautilus. Cephalopod external shells were heavy, unstreamlined, and imploded below critical depths. Hence ancient cephalopods were restricted to narrow coastal seas and were likely slower, perhaps more scavenging than predatory. The success of modern cephalopods is in no small part due to the evolution of alternative solutions to buoyancy allowing them to reduce their shell.

SHELL REDUCTION & BUOYANCY COMPENSATION

In all modern cephalopods apart from nautiluses the shell is internalized. The cuttlebone of a cuttlefish is an internal calcareous shell that has become flattened, with the gas-filled chambered section—the phragmocone—spread along its ventral surface as a series of obliquely oriented leaflets.

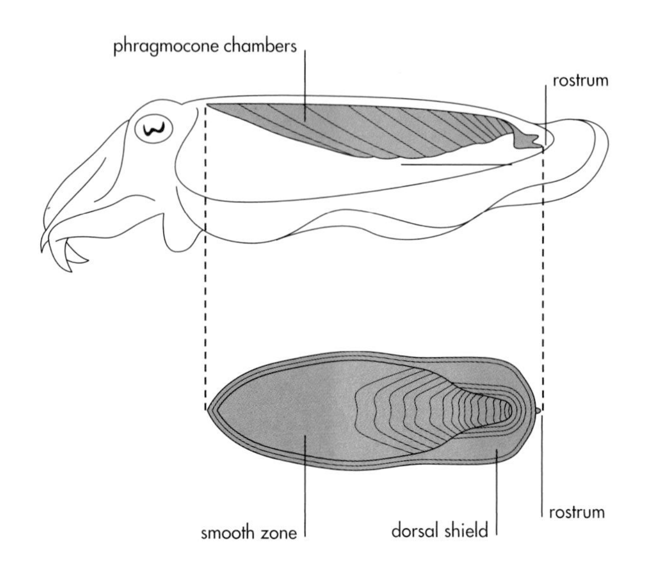

The unusual oceanic ram's horn squid, *Spirula spirula*, also has an internal, chambered, calcareous shell that is coiled the opposite way to nautilid shells. By retaining their gas-filled chambered shell, cuttlefishes and *Spirula* solve buoyancy in a similar fashion to ancient cephalopods.

In inshore and oceanic squids, some bobtail squids, and the vampire "squid," the internal shell is further reduced to a chitinous structure shaped somewhat like a feather along the dorsal midline of the mantle. This structure is called a gladius or pen (as in quill pen). Some bobtail squids and all bottletail squids have no shell. Constant fin swimming to overcome gravity is used in many squid species, although squids and pelagic octopods have evolved other solutions too.

A few squids, for example *Grimalditeuthis bonplandi*, have a buoyant structure on the tail acting as a flotation device. Other squids concentrate ammonium ions in their tissues, or, in the case of family Cranchiidae, in an internal body cavity, the coelom. The resulting solution of ammonium chloride is lighter than seawater and increases the animals' buoyancy.

Opposite The structure of the cuttlebone and its position in a cuttlefish.

Top right A shell of the oceanic ram's horn squid, *Spirula spirula*.

Middle right The ram's horn squid, *Spirula spirula*. The planispiral shell (top right) sits internally at the end of the mantle.

Bottom right The deep-sea squid, *Grimalditeuthis bonplandi*, in the deep waters of Monterey canyon.

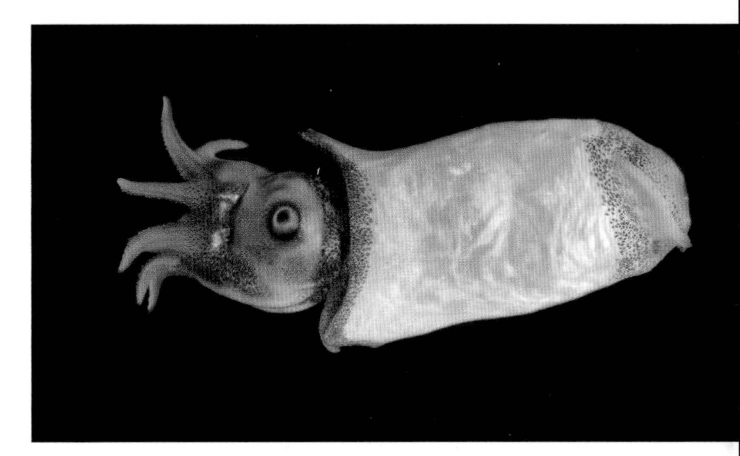

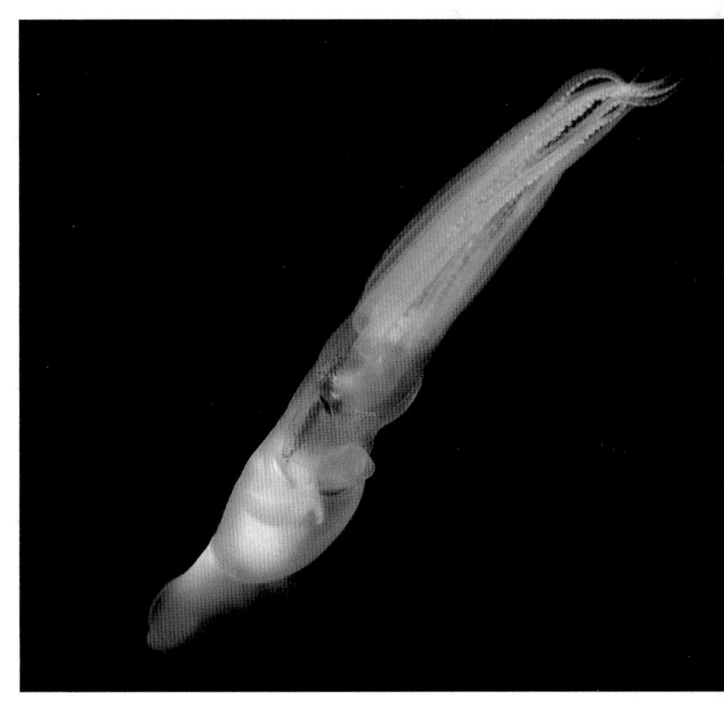

In the deep-sea cirrate, or finned, octopods, the internal shell forms a substantial cartilaginous structure consisting of two lateral projections to which the cartilage and muscles of the fins attach. Slow energy-efficient swimming using their fins may provide sufficient lift to compensate for the lack of phragmocone. The shell has been further reduced in incirrate octopods to a pair of small rod-like stylets embedded dorsolaterally in the mantle musculature where the funnel retractor muscles attach to the internal wall. For octopods living on the seafloor, no compensatory buoyancy mechanism is required. A small group of pelagic octopods replace sulfate ions with lighter chloride ions to reduce their density.

The most remarkable solution is found in females of three genera of argonautoid octopods, which have evolved true swim bladders, probably from a pouch of the digestive system. A female of the fourth genus, *Argonauta* (see species profile p140), instead uses her egg case, which has completely different evolutionary and developmental origins from the shell. She traps small bubbles of air in her egg case and can adjust the air volume to control her buoyancy.

OCTOPODIFORMES

ONE OF TWO MAIN LINEAGES OF LIVING coleoids, Octopodiformes includes the vampire "squid" and all types of octopod, including benthic octopuses such as the common octopus and the Caribbean reef octopus, the deep-sea "dumbo" octopods and flapjack devilfishes, and unusual groups such as the argonautoids. All octopodiform species have eight arms, although vampire "squid" also have a pair of retractile filaments that evolved from two additional arms in an ancestor.

THE VAMPIRE "SQUID"—A LIVING FOSSIL OCTOPODIFORM

The vampire "squid" (species profile page 90), is more closely related to octopods than squids. It represents a unique lineage of Octopodiformes and, to avoid confusion, we refer to them simply as "vampires." The single living species, *Vampyroteuthis infernalis*, inhabits mesopelagic to bathypelagic depths where it gently flaps its fins, swimming in search of food and detritus. Two sticky filaments trailing from its arm crown capture small food particles that the arms pass to the mouth. Vampyromorphs diverged from coleoid ancestors at the start of the Jurassic and were more diverse in earlier times. Nonetheless, exceptional fossils, for example *Vampyronassa*, show these creatures to be little changed in more than 150 million years. The vampire is an ancient relict of Jurassic seas—a living fossil.

The development of vampires is fascinating. Hatchlings have a single posterior pair of fins but, as the hatchling grows, these are resorbed and replaced by a more anterior pair. Just briefly, the hatchling has two pairs of fins—a feature unique among modern cephalopods. Among fossils, two pairs of fins have been reported in *Trachyteuthis*, a possible relative of vampyromorphs from the Upper Jurassic. Maybe what we see in vampire hatchlings is a hangover from an earlier evolutionary condition?

Below left The vampire "squid," *Vampyroteuthis infernalis*.

Below A flapjack devilfish, one of around sixty species of cirrate octopod found in deep waters.

Opposite Evolutionary tree of Octopodiformes. *Proteroctopus* appears with a question mark because paleontologists are not yet certain on which lineage it belongs.

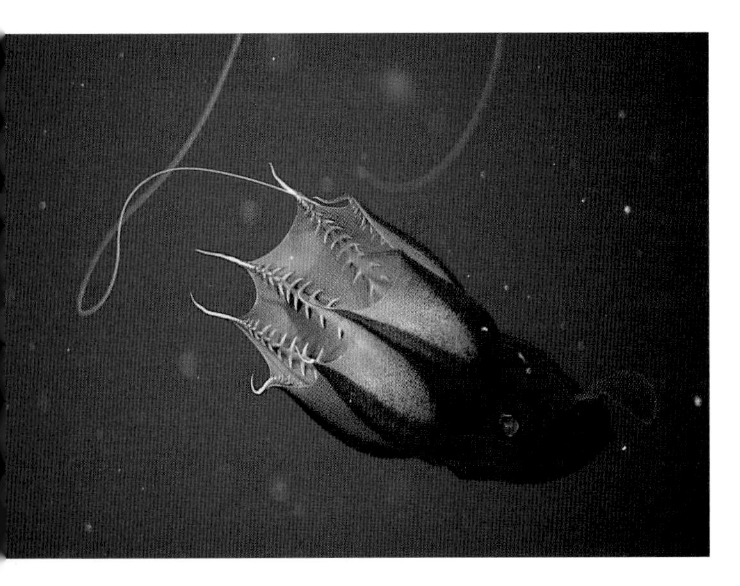

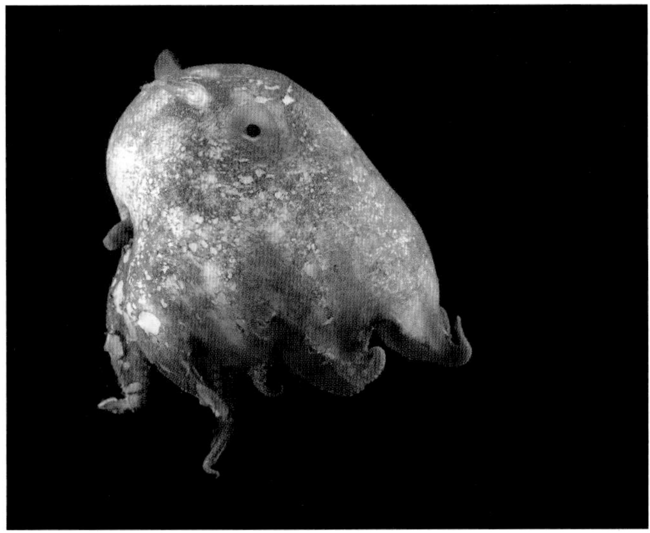

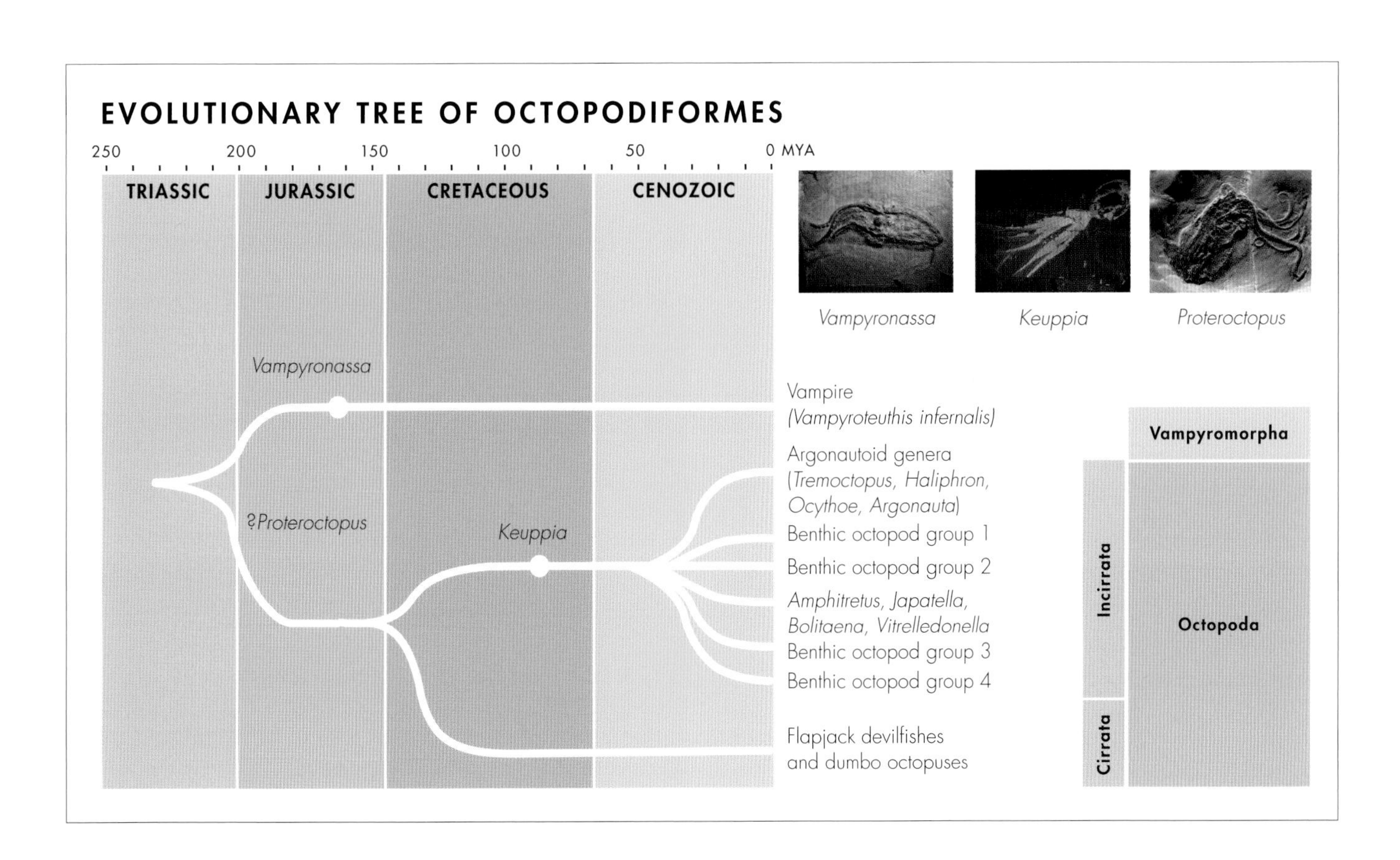

EVOLUTION OF OCTOPODS

The unique feeding method of vampyromorphs is thought to have been important in the development of a benthic lifestyle in incirrate octopods. Vampire brains have highly developed inferior frontal lobes for processing chemotactile information from the arms. This ability in their ancestors likely made the early octopods pre-adapted for exploring the seafloor with their arms splayed and their oral surfaces facing downward.

In contrast to the single living species of Vampyromorpha, there are more than 300 named species of Octopoda, with many more yet to be discovered and described. They are divided into two main groups, the cirrates (or Cirrata), which have fins for swimming and tend to live in the ocean depths, and the incirrates (or Incirrata), which have no fins and live mostly on the seafloor and are commonly encountered in temperate and tropical reef habitats worldwide.

Although well-preserved octopod fossils are rare, another Lagerstätte in Lebanon, at Hajula, has yielded stunning fossil octopods: one finned, *Palaeoctopus*, and two unfinned, *Styletoctopus* and *Keuppia*, providing evidence that octopods very similar to those we see today have existed for at least 100 million years. However, it is uncertain when the first octopuses appeared because very old fossils once thought to be octopods, such as *Pohlsepia* and *Proteroctopus*, may belong to other lineages.

CIRRATA: THE DUMBO OCTOPODS & FLAPJACK DEVILFISHES

Encompassing more than 60 species, cirrate octopods inhabit deep waters on continental slopes and abyssal plains worldwide. They are named for cirri, the small, finger-like projections that alternate with suckers along the length of their arms. Cirri are present on the arms of vampires too, but not on the arms of incirrate octopods. Cirrates are also recognizable by their deep webs, the paired lateral fins that they use to swim, and by the distinct internal cartilaginous shell that supports the fins. Cirrates produce a few hundred to a few thousand very large eggs, which they lay singly, attached to corals and other objects on the seafloor. Based on their size and the water temperature in which they are laid, these eggs probably develop for 2–3 years before hatching. This slow life cycle makes cirrates slow to replenish their populations. The IUCN considers five species of cirrate octopods to be threatened, all in areas where deep-sea fisheries, which target other species but take cirrate octopods as by-catch, are active. *Opisthoteuthis chathamensis*, found off the east coast of the North Island of New Zealand, has been categorized as Critically Endangered, and has not been seen for nearly 20 years. It may even be extinct.

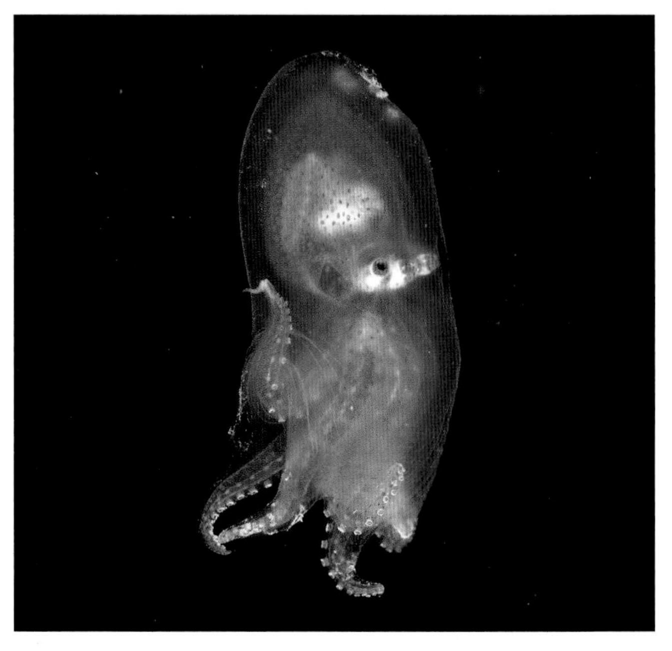

Top *Grimpoteuthis discoveryi* viewed from below. Cirri and suckers alternate up the arms.

Above The small transparent incirrate *Amphitretus pelagicus* floating in the water column. Note its tubular eyes.

INCIRRATA: A DIVERSITY OF FORMS

These are the most familiar octopods that include the common shallow-water octopuses frequently encountered by divers, and encompass a cluster of families. They are defined by their lack of cirri and absence of fins. Most incirrate octopods are benthic: they live on the seafloor, exploring their surroundings with their flexible and sensory arms. Found from the icy waters of polar seas to the balmy waters of the tropics, they inhabit a massive range of depths from the intertidal to more than 13,000 ft (4,000 m) deep. New explorations with cameras on remotely operated vehicles are revealing new benthic species in ever more extreme habitats. Many have become masters of disguise, their complex nervous systems controlling muscle fibers that can show or hide colorful chromatophores and alter their skin texture, allowing them to blend seamlessly into their seascape. Incirrates no longer require a shell for either buoyancy or support, and their complete shell reduction to, at most, two small stylets, allows them to squeeze through very small gaps and make their homes in cryptic habitats such as tropical reefs and rocky shores.

Above The transparent pelagic incirrate *Vitreledonella richardi* can be found in tropical and subtropical waters.

A SECONDARILY PELAGIC LIFESTYLE

A handful of incirrate octopods have redeveloped a pelagic lifestyle. *Amphitretus pelagicus* and four closely related species appear to have become pelagic through neoteny—the retention of larval features into adulthood. Many benthic octopods lay hundreds to thousands of tiny eggs whose hatchlings are buoyant and drift in the plankton. *Amphitretus* has retained this buoyancy into adulthood, replacing sulfate ions in its tissues with chloride. It is gelatinous and transparent, its shell completely absent. As remotely operated vehicles capture images of these enigmatic species, we are slowly learning more about their remarkable adaptations to reproduction in the pelagic realm. Benthic octopods attach their eggs to rocks and other structures on the seafloor: the pelagic *Bolitaena pygmaea* and *Japatella diaphana* brood their eggs in their arms while *Vitreledonella richardi* broods its eggs within its mantle cavity. The pelagic octopods include the only known incirrate species with a light organ, perhaps used to attract a mate in its vast three-dimensional domain.

Above A female of the genus *Argonauta* residing in the egg case which she secretes with special glands on her arms.

Opposite *Ocythoe tuberculata*, the football octopus, which has a true swim bladder.

The four argonautoid genera have adapted differently to the pelagic realm and display some of the most unusual evolutionary adaptations seen in octopods. They have developed unique solutions to buoyancy (see The Buoyancy Conundrum page 64), extreme sexual dimorphism (see On Dwarfs & Giants page 80), and have resolved the difficulties of pelagic reproduction and egg care with style. The male hectocotylus—the organ for transferring sperm to the females—is detachable in argonautoid families. The female stores the male hectocotylus and the sperm packets associated with it until she is ready to fertilize her eggs. Female argonautoids also have remarkable methods for brooding their eggs. *Haliphron atlanticus* appears simply to hold her eggs in her web.

Ocythoe tuberculata is one of just two cephalopods known to be ovoviviparous (gives birth to live young). A *Tremoctopus* female uses her dorsal arms to secrete a rod-like structure to which she attaches her eggs, which she then carries with her until hatching. A female *Argonauta* uses secretory glands similar to those of *Tremoctopus* to secrete a shell-like egg case within which both she and her eggs reside.

DECAPODIFORMES

DECAPODIFORMES, WITH THEIR TEN arms (which include two specialized as tentacles), are remarkably diverse. They vary in size from tiny pygmy squids to giant and colossal squids and occur from the shallowest coastal areas to the depths of the oceans. The seven modern lineages began to diverge in the Jurassic and, over the next 150 million years, further developed their skin signaling systems and became fierce predators.

THE FOSSIL RECORD

Sadly, Lagerstätten have not yet yielded incredibly well preserved fossils of close relatives of modern squids. Squids—recognized by the "pen" or "gladius," the very reduced shell that supports the length of the mantle—are barely known from the fossil record. Their gladii have a high organic content and do not fossilize well. Putative fossil squids are limited to

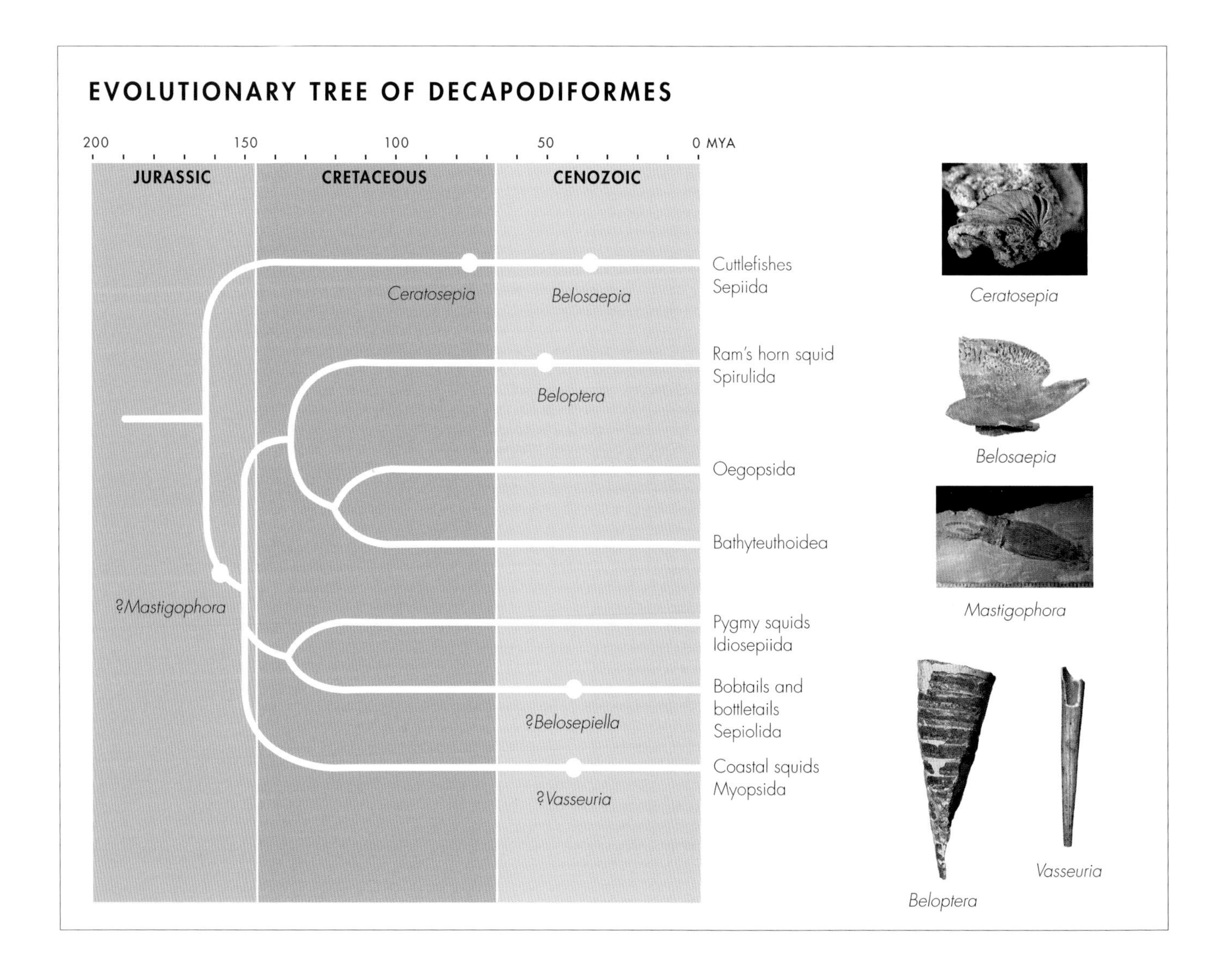

Opposite In this evolutionary tree of Decapodiformes, question marks indicate where paleontologists cannot be certain of the lineage.

Left A cuttlefish with its tentacles partially extended in a hunting strike.

the enigmatic *Vasseuria*, *Mastigophora*, and *Belosepiella*, which could be ancestral squids or bobtails but could equally represent different lineages. For the most part, the fossil record of Decapodiformes is limited to those groups with more extensive shells. Spirulida—the single modern representative of which, the ram's horn squid *Spirula spirula*, has an internal coiled chambered shell or phragmocone—are known from the late Jurassic onwards. However, the shells of spirulid fossils from the Jurassic and Cretaceous were not coiled. It was not until the Paleozoic that coiled spirulid shells appeared.

Cuttlefishes, with their dorsoventrally flattened chambered cuttlebone, are known with certainty from the Late Cretaceous, 80 million years or so after spirulids. But there are earlier fossils—*Voltzia* from the Upper Jurassic and *Actinosepia* from the Cretaceous—that may also be in the sepiid lineage.

CUTTLEFISHES

Cuttlefishes are familiar from their chalky white cuttlebone that is often found washed up on beaches. This, their internal chambered shell, provides them with buoyancy in life. More than 100 species of cuttlefishes exist today, living mostly in the shallow waters of continental shelves, although some species may reach depths of around 2,000 ft (600 m). They are absent from polar regions and the Americas, their existence in these waters having been prevented by past or present low temperatures. Fossil cuttlebones tend to be found in strata that were once shallow warm seas, suggesting that their habitat has changed little over geological time.

Cuttlefishes have evolved an incredible chromatophore system. From hatching, the lobes of the brain responsible for chromatophore control are highly developed, and chromatophores are densely packed on the skin surface. As adults, cuttlefishes display some of the most complex signaling patterns known in the animal kingdom.

Like other ten-armed cephalopods, two of the cuttlefish's arms are highly specialized as tentacles that can shoot out and grab prey. Even among cephalopods, cuttlefish tentacles are highly evolved. Elastic and retractile, controlled by an animal with highly evolved brain and vision, their tentacles are deadly and sophisticated hunting weapons.

SQUIDS OF THE OPEN OCEAN

Spirula spirula, the ram's horn squid, is found in mid-water at the edge of continental shelves where the seafloor depth descends more rapidly, and above similar slope habitats around islands, seamounts, and submarine volcanoes. Because spirulids have been known from so far back in geological history, the ram's horn squid has sometimes been described as a living fossil. However, in reality, the lineage has evolved substantially over time, with the modern species bearing less resemblance to its Jurassic ancestors than, for example, the vampire does to its ancestors.

The exact relationships among decapodiform lineages are not known, but DNA analyses suggest that the two lineages of oceanic squids and the ram's horn squid evolved from a common ancestor. This suggests a single invasion of the pelagic realm by this group and that the cornea, which is lacking in all three lineages, was lost in an early pelagic ancestor of the group. Corneas protect the eyes and it is likely that protection was much less important in an open water environment than in a silty near-bottom (demersal) environment, such as that favored by cuttlefishes.

Oceanic squids in the order Oegopsida are diverse and abundant. They have evolved the classic streamlined body and some are powerful and fast swimmers, excelling at their predatory lifestyle. But the families of oegopsid squid differ enormously, and although some may be muscular, others have

THE SEVEN LINEAGES OF MODERN DECAPODIFORMES

Lineage	Common name	Number of living species	Shell	Habitat	Cornea	
Sepiida	Cuttlefishes	~120	Chambered	Coastal demersal	Yes	
Spirulida	Ram's horn squid	1	Chambered	Oceanic pelagic	No	
Oegopsida	Oceanic squids	~230	Gladius	Oceanic pelagic	No	
Bathyteuthoidea	Oceanic squids	~6	Gladius	Oceanic pelagic	No	
Idiosepiida	Pygmy squids	~6	Gladius	Coastal demersal	Yes	
Sepiolida	Bobtail squids or bobtails (family Sepiolidae) and bottletail squids or bottletails (family Sepiadariidae)	~70	Reduced gladius	Coastal demersal /pelagic	Yes	
Myopsida	Inshore squids	~50	Gladius	Coastal demersal	Yes	

evolved an almost transparent body that provides camouflage while the evolution of photophores that produce light through biochemical reactions peaks in mesopelagic species such as *Histioteuthis*. Their adaptation to their pelagic environment is seen in their egg masses, which are neutrally buoyant and may float free in the ocean, in contrast to the stalked egg masses that are attached to some form of substrate by most inshore cephalopods.

INSHORE SQUIDS

Squid species in the order Myopsida may live in extremely shallow coastal waters. Some species are found on the edge of the continental shelves but they do not reach into the open ocean and they tend to live close to the seafloor, again with a cornea to protect their eyes. Like the order Oegopsida, they retain the classic streamlined body shape with pointed tail, but they vary enormously in size and can be tiny. This is a group in which bacterial photophores have evolved (see Chapter 4). The ability to produce light in the deeper darker waters has clearly been selected for over evolutionary time.

PYGMY, BOBTAIL & BOTTLETAIL "SQUIDS"

Pygmy squids, bobtail squids, and bottletail squids appear to be closely related to one another, united by morphological characteristics such as the form of the gladius, a rounded end to the mantle, and muscles attaching the viscera to the mantle. Strictly, they are not squids, and we try to avoid confusion throughout this book by referring to them instead as idiosepiids, bobtails, and bottletails.

Idiosepiids are the smallest of all living cephalopods, maturing when their mantles are barely 5⁄12 in (1 cm) long. They live in shallow coastal beds of seagrass in tropical habitats, and have evolved a curious adhesive organ that allows them to glue themselves to blades of foliage. Uninhibited by their tiny size, idiosepiids are ferocious predators, taking shrimps almost as large as themselves as prey.

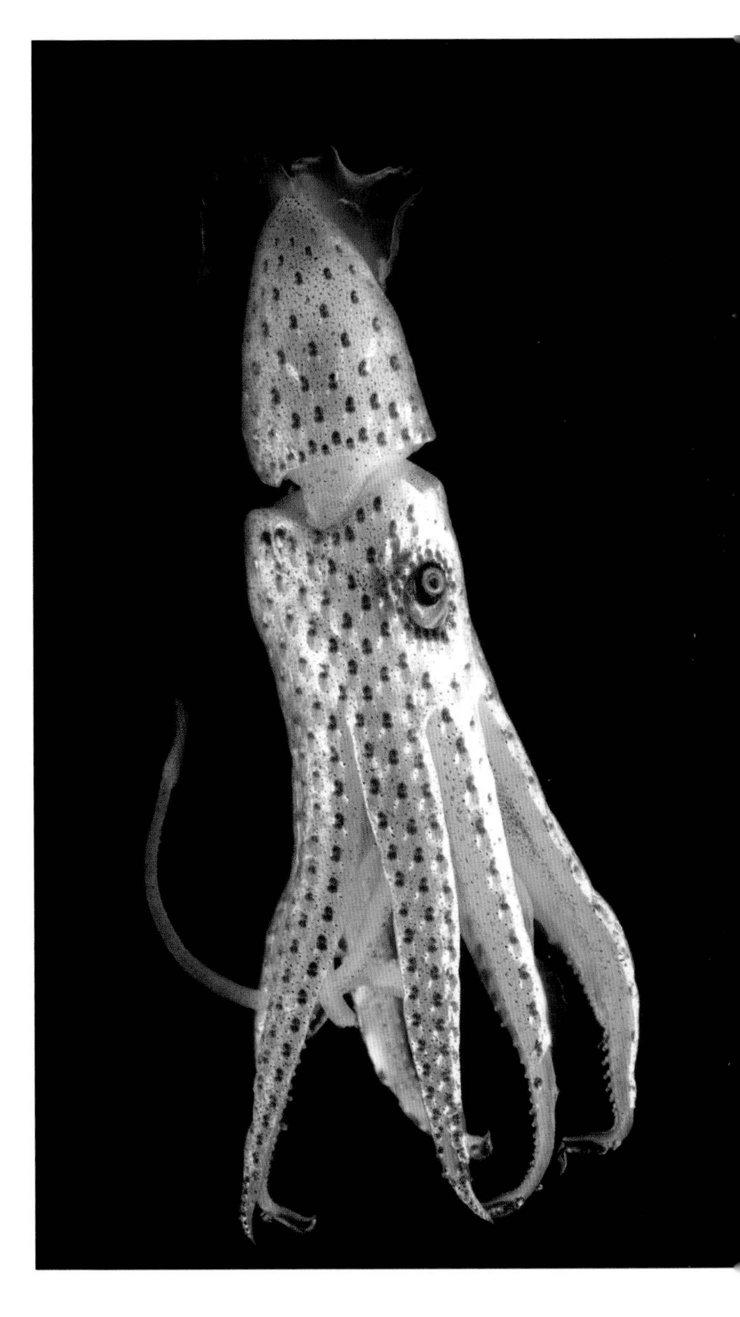

Above The entire body of the jewel squid, genus *Histioteuthis*, is covered in photophores.

Bobtails (family Sepiolidae) and bottletails (family Sepiadariidae) mostly live on or close to the seafloor mostly in coastal waters, but one group of bobtails (subfamily Heteroteuthinae) has evolved a pelagic or at least semi-pelagic lifestyle and may be found in the open ocean. Unlike the open ocean squids discussed previously, heteroteuthins retain the cornea, suggesting perhaps that their evolution into pelagic waters has been more recent. This is another group in which bacterial photophores have evolved. The light-producing symbiosis has been extensively studied in the Hawaiian bobtail, *Euprymna scolopes*, and its relationship with the bacterium *Vibrio fischeri* has become one of the best understood models of animal–bacteria symbioses.

DNA analyses place the split between idiosepiids and bobtails in the early Cretaceous, suggesting that, while evolution within groups may be recent or ongoing, the main lineages themselves diverged early from one another and have existed as independent lineages for close to 150 million years.

Below The Hawaiian bobtail, *Euprymna scolopes,* inhabits shallow coastal waters.

Opposite top Inshore squids of the genus *Sepioteuthis* are commonly known as reef squids because of their association with coral reefs in tropical and subtropical waters. But they are also found in areas where reefs are absent.

Opposite bottom The outline of the Japanese firefly squid, *Watasenia scintillans,* picked out by the glow of its photophores.

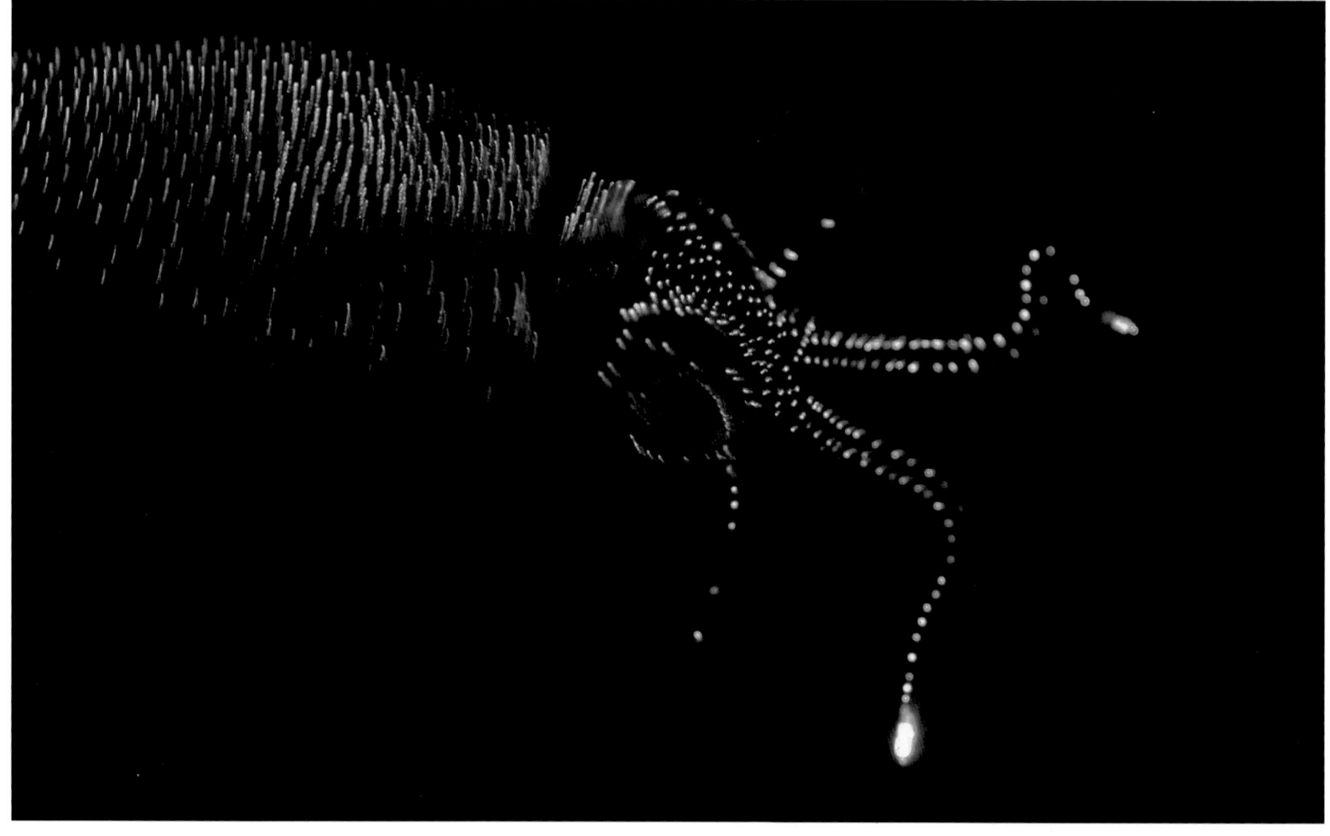

CEPHALOPODS & FISHES: CONVERGENT EVOLUTION

CEPHALOPODS AND FISHES HAVE EVOLVED in parallel since they arose over 500 million years ago. Occupying similar niches in similar habitats, they have been subject to similar, but changing, selective pressures over more than 500 million years. From the smallest to the largest species, their features are often convergent—originating differently, but resulting in similarity. Both groups have been highly successful, competition not having caused the demise of one or other.

THE RISE OF FISHES

Like cephalopods, the first fishes appeared in the Cambrian. Heavily armored fishes were prevalent in the Ordovician, at a time when cephalopods were developing their shells and gaining greater buoyancy. The soft inner tissues of both groups were well, if differently, protected: an essential defense against the giant sea scorpions—the eurypterids—that flourished in this period. The Devonian, heralded as "The Age of Fishes," because of the huge variety of fish species that evolved at that time, also saw the rise of two massively successful cephalopod groups, the ammonites and the nautiloids. The great Permian extinction reduced the fishes significantly, but those cartilaginous and bony fishes that survived this event steadily diversified throughout the Triassic, Jurassic, and Cretaceous, before radiating massively in the Cenozoic. Coleoid cephalopods similarly diversified from the Triassic to Jurassic and, although not diversifying to match the huge number of marine fish species (18,000 versus about 1,000 cephalopod species), octopods, cuttlefishes, and squids have diversified to occupy nearly every marine habitat.

Top left Sardines swarming in a bait ball in response to predators.

Left A shoal of squid off Cuba. Shoaling is common to many squids and fishes.

SIMILAR NICHES

Fishes and cephalopods occupy similar niches wherever they are found. Many benthic octopods are territorial hunters. So are reef-dwelling groupers and scorpionfishes. Cuttlefishes on sandy bottoms have been compared to the flatfishes in the same environment. Inshore squids shoal like many coastal fishes— Atlantic mackerel, for example—as a defense against predators and to maximize foraging success. Many oceanic squids can be compared to the epipelagic bony fishes: both groups are fast, voracious predators. Squids and fishes alike have evolved streamlined bodies that are extremely hydrodynamically efficient.

Like cephalopods, fish sizes range from tiny to massive. The smallest fish, *Schindleria brevipinguis*, the stout infantfish from Australia's Great Barrier Reef, grows to just $^{5}/_{16}$ in (8 mm) long. The whale shark can reach 40 ft (12 m) in length. *Dunkleosteus*, a Devonian fish, reached 20 ft (6 m). These sizes are comparable to the small idiosepiids and the largest living squids, the giant and colossal squids, and the largest prehistoric cephalopods, such as *Cameroceras*. Hence, cephalopods and fishes have been and are competing across similar size ranges.

CONVERGENT EVOLUTION

Living in the same environment, at the same sizes, has meant that cephalopods and fishes have been subject to similar selective pressures. They also prey on each other, in addition to competing for similar resources. The forces that have molded evolution in the fishes, have molded evolution in the cephalopods. As one group advanced, the selective pressure on the other group would have increased, potentially driving evolution. Certainly, the molluscan body plan has become highly adapted in cephalopods, with locomotory and sensory mechanisms reaching levels only otherwise seen in vertebrates.

Above The blue-spotted rock cod, *Cephalopholis miniata*, is a territorial hunter on the reefs of the Indo Pacific.

These convergent successes are apparent in the highly developed cephalopod eye, the myriad of adaptations to achieve buoyancy, the cephalopod jet propulsion system, the development of chromatophores and photophores to facilitate complex camouflage and signaling, and not least in the cephalopod brain, which is the most developed of any invertebrate. In fact, cephalopod evolution has arguably solved some problems better than fish evolution; for example, a cephalopod that can orient its funnel in any direction, can move backward as fast as it can forward. Also, cephalopods can catch and eat a larger range of prey sizes than can bony fishes, which are limited by their mouth size. However, molluscan heritage has drawbacks: cephalopod locomotion and respiration are not as energy efficient as in vertebrates. It could be that this physiological disadvantage has led to selection for rapid growth and short life histories in cephalopods, reducing direct competition with and predation by fishes and allowing survival of both groups.

ON DWARFS & GIANTS

CEPHALOPODS HAVE PERHAPS A GREATER range of adult sizes than any other animal group. The largest species at maturity are more than 200 times the length of the smallest species at maturity. It is a pattern seen in octopods and squids alike, from *Enteroctopus dofleini*, the giant Pacific octopus, to *Octopus wolfi*, the smallest known species; from the giant squid *Architeuthis* to the thumbnail-sized *Idiosepius*. Their success at such a range of body sizes is testament to their extraordinary evolutionary adaptations.

PREHISTORIC GIANTS

Most people have heard of giant squid, but fewer know that there were giant cephalopods in prehistoric seas. One of the earliest and largest was *Cameroceras*, a straight-shelled nautiloid monster of the Ordovician. Reaching perhaps more than 20 ft (6 m) in length, its phragmocone barely compensated for the weight of the shell itself, with large individuals probably resting near the seafloor and relying on ambush to take their prey. Although possibly the largest, it was not the only enormous nautiloid of this time, the prevailing conditions clearly being conducive to the evolution of massive size.

Ammonites grew similarly large. The largest known, *Parapuzosia seppendradensis*, estimated to reach diameters of 8 ft (2.5 m), thrived in the Late Cretaceous. *Parapuzosia* was likely as unmaneuverable as *Cameroceras*, with an estimated weight of about a ton and a half, and probably a similar lifestyle.

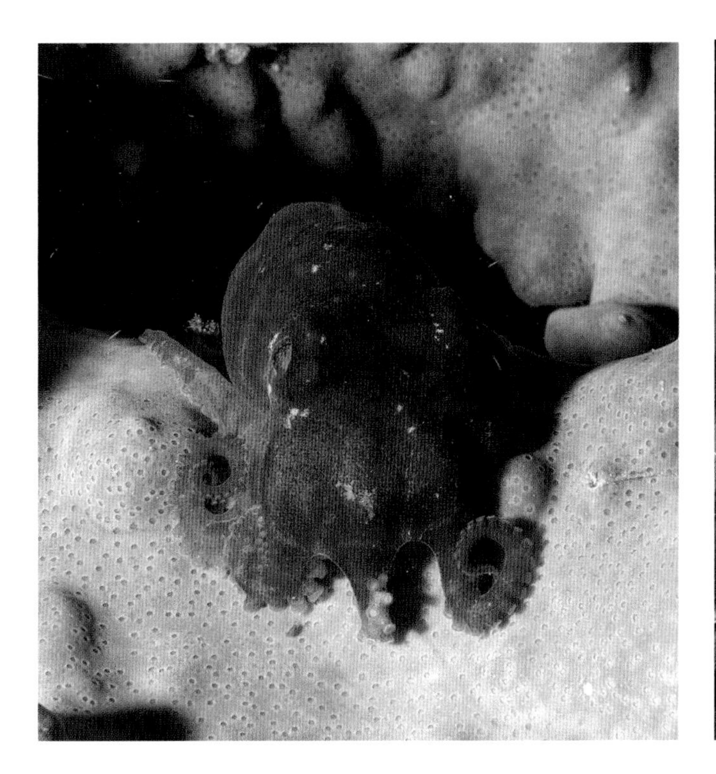

The Jurassic belemnite *Megateuthis* had greater maneuverability. Like many belemnites, it is known from the rostrum, or guard, of the internal shell. With rostra reaching about half a yard (just under half a meter) in length, the mantle was likely close to 6½ ft (2 m) in length and the whole animal from tail to arm tips might even have reached 16½ ft (5 m). *Tusoteuthis*, a vampyromorph from the Cretaceous, was probably similarly sized.

In contrast to the above species that are known from their shells, *Yezoteuthis*, a coleoid also from the Cretaceous, is known from its fossilized jaws or beak. These are similar in size to those of the giant squid, suggesting parity in length with *Megateuthis* and *Yezoteuthis*.

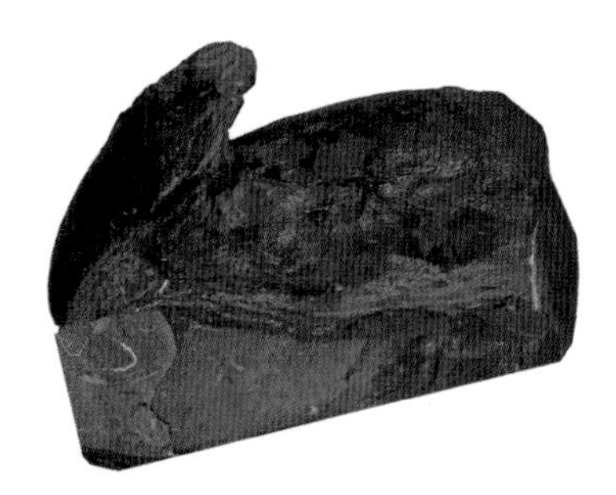

Opposite The giant Pacific octopus, *Enteroctopus dofleini*, which inhabits coastal shelves of the western US and Canada.

Top left The mantle of an adult of *Octopus wolfi*, the star-sucker pygmy octopus, measures less than 1 in (2.5 cm).

Top right Visitors admire the huge ammonite *Parapuzosia seppendradensis* in the Westphalian Museum of Natural History.

Above The very large beak of the extinct *Yezoteuthis*.

Above A live giant squid, *Architeuthis dux*, hauled to the surface on a baited hook in Japan.

RECENT GIANTS

The giant squid, *Architeuthis dux*, has similar dimensions to Jurassic and Cretaceous giant coleoids. A fully-grown adult has a mantle more than 6½ ft (2 m) long, with its head and arms extending its length to around 16½ ft (5 m). However, with its long, very elastic tentacles fully extended, its length from tip to tail can be as much as 40 ft (12.2 m). It can weigh up to a ton (900 kg). Again, it is not the sole giant. The colossal squid, *Mesonychoteuthis hamiltoni*, is probably heavier, if shorter in overall length, and there are at least eleven squid species and four octopods that are larger than us—that is, having a total length exceeding 6½ ft (2 m).

IDIOSEPIIDS & OCTOPODS

Cephalopods can be tiny too! The smallest belemnites were about 4 in (10 cm) long. The earliest known cephalopod, *Plectronoceras*, was little more than 5⁄12 in (1 cm) long. The largest mature female of the idiosepiid *Idiosepius thailandicus* has a mantle just 5⁄12 in (1 cm) long, and the male is even smaller. *Octopus wolfi*, the star-sucker pygmy octopus, grows to 7⁄12 in (1.5 cm) mantle length, its arms extending its total length to about 2 in (5 cm). The Antarctic pygmy octopus, *Bathypurpurata profunda*, manages to store eggs 3⁄16 in (4 mm) long in a body just ⅞ in (2.3 cm) long. To be so successful in so many different niches, as cephalopods have been, requires remarkable plasticity. Their growth, their behaviors, and their reproductive strategies contribute hugely to their success.

DWARF MALES

Sexual dimorphism—males and females of differing shapes and sizes—is common in many animals. It is known even in ammonites, discernible from fossilized shells. In species where males compete for females, the males tend to be larger; in other species, the females may be larger to increase investment in egg production. But dimorphism is taken to its extreme in the argonautoid genera. Males may be as little as 5 percent of female size and mature at just 5⁄12 in (1 cm) mantle length. The dwarf males of the different genera have evolved remarkable defensive behaviors, which include carrying stinging cells from jellyfishes and hiding inside salps.

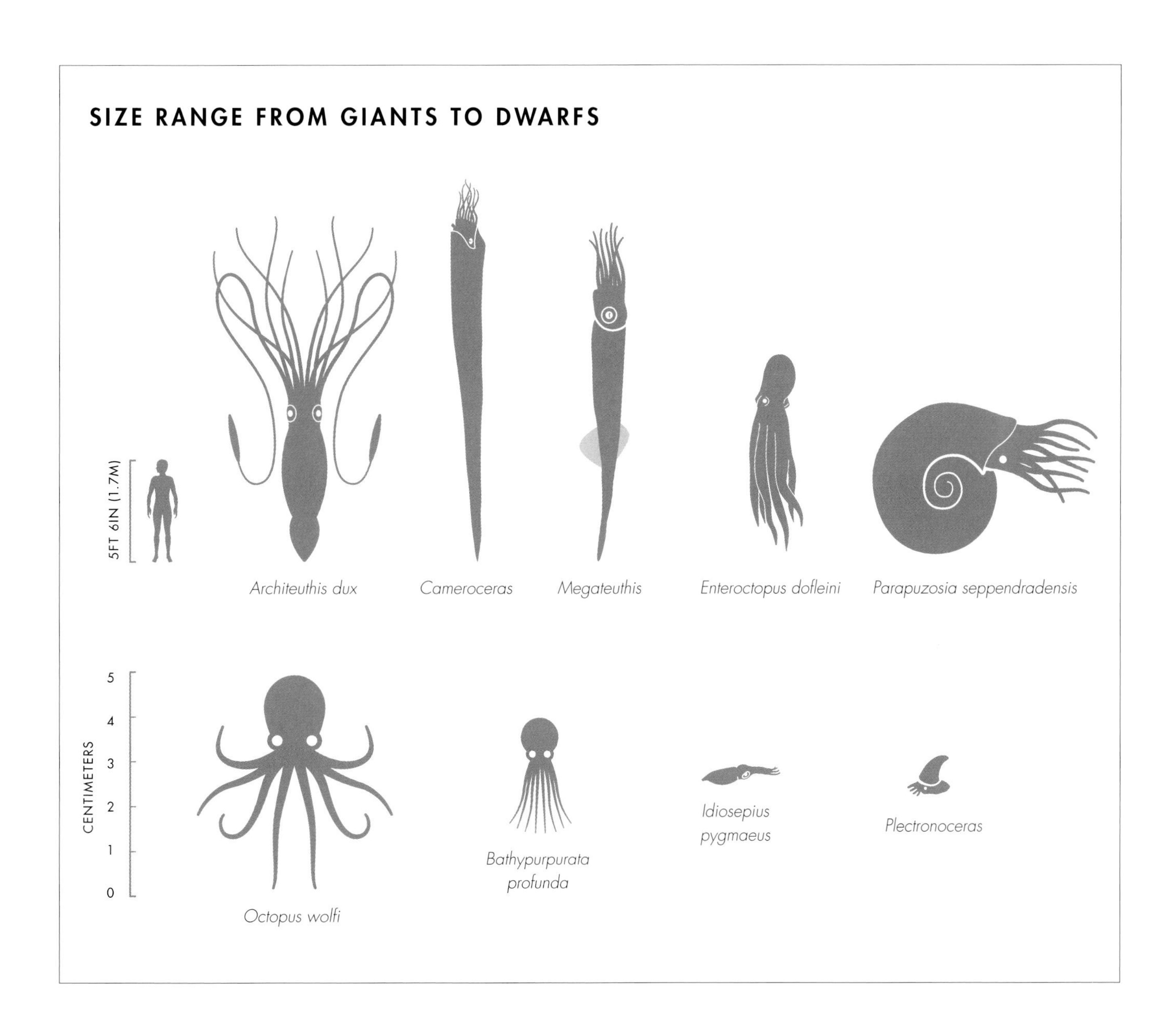
SIZE RANGE FROM GIANTS TO DWARFS
5FT 6IN (1.7M)
Architeuthis dux
Cameroceras
Megateuthis
Enteroctopus dofleini
Parapuzosia seppendradensis
CENTIMETERS
5
4
3
2
1
0
Octopus wolfi
Bathypurpurata profunda
Idiosepius pygmaeus
Plectronoceras

EVOLUTION & CLIMATE CHANGE

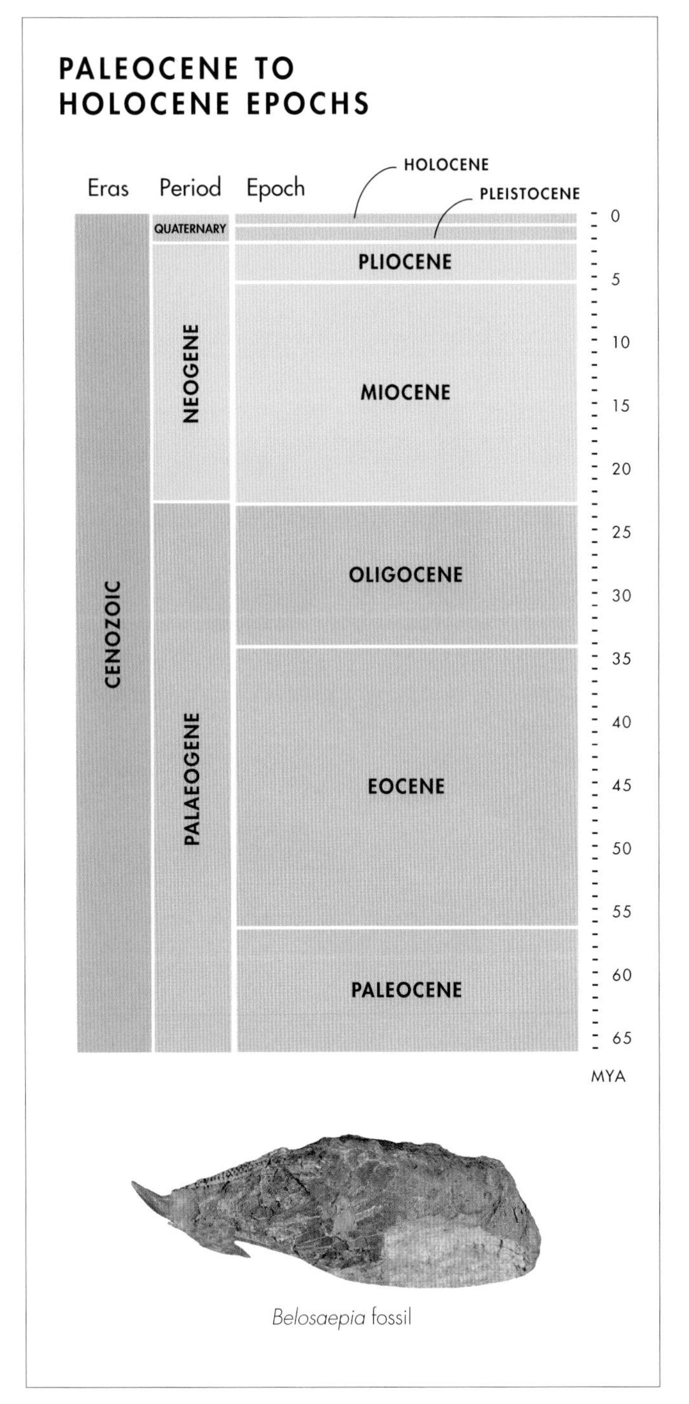

Belosaepia fossil

PATTERNS OF CEPHALOPOD DISTRIBUTION through geological time reveal much about their evolution. Some groups, such as the cuttlefishes, have left an extensive fossil record and we can trace their demise and their subsequent recovery. In other cephalopods, the fossil record may be poorer, but their evolutionary history has left its mark on their DNA. When we compare these histories with climate records we find that climate can be a remarkable driver of cephalopod evolution.

ANTARCTICA: CRADLE OF EVOLUTION

Since the Late Eocene, climate change has been massively influential on cephalopod diversity and Antarctica has been at the heart of this. Southern seas began to cool at the start of the Oligocene, about 34 million years ago, as the other continents moved northwards away from Antarctica, isolating it, and allowing the Antarctic Circumpolar Current to flow. These colder conditions initiated the diversification of a unique lineage of Antarctic octopods. Twenty million years later, during the mid-Miocene, the Antarctic ice sheets expanded dramatically and the Southern Ocean cooled further. Sinking of dense cold water in Antarctica stimulated a global current system and spread frigid Antarctic water over the floor of other oceans. This new habitat was seized by the newly successful Antarctic octopods, and they slowly moved northward. Many of the octopods found in deep oceans today evolved from this Antarctic stock.

A NEW WORLD DEVOID OF CUTTLEFISHES

The cooling temperatures of the Oligocene had other impacts. In the Eocene, an epoch marked by mostly tropical conditions globally, the dominant cuttlefish genus *Belosaepia* could be found off northwest Europe, off landmasses that have since become Turkey and India, and off southern North America. Just ten or so million years later, the much colder seas of the Oligocene led to the extinction of cuttlefishes from the New World and many other areas. The genus *Sepia*, which holds

Above left The longfin inshore squid, *Doryteuthis pealeii*, is fished off the eastern coast of the US.

Above The bigfin reef squid, *Sepioteuthis lessoniana*, gets its common name from its large oval fin that extends all the way along the mantle to the anterior end of the squid.

95 percent of all modern cuttlefishes, probably spread subsequently from the Mediterranean Sea to the developing Indian Ocean and began a radiation through Southeast Asia, where it is hugely diverse and abundant today. But why did cuttlefishes not spread back to the Americas? The answer lies in their coastal and temperate to tropical habitats and lack of planktonic paralarvae. Confined to shallow shelf and slope seas, the only routes by which cuttlefishes could re-colonize the New World were via the Bering Straits or from Europe to Greenland via the Faroe Islands and Iceland. Since the Oligocene, neither of these northern routes has been warm enough, so the New World is devoid of cuttlefishes.

SEPIOTEUTHIS, A TETHYAN RELIC

Like cuttlefishes, inshore squids live solely in coastal regions. There are only about 50 species worldwide, and nearly all genera are confined to very specific locations; for example, the squid genus *Doryteuthis* is only found around the Americas. But one genus, *Sepioteuthis*, bucks this trend with a distribution that extends from the southeast Mediterranean through the Indian Ocean to the Indo-West Pacific, east to Hawaii and even into the Caribbean.

The clue to its almost worldwide distribution lies in its age. DNA studies have shown it to branch off first on the inshore-squid tree of life, suggesting that it is older than the other genera. The best explanation for its wide distribution is that it is a Tethyan Relic. The oldest fossil in this group, *Vasseuria*, dates from the Eocene, when not only were temperatures warmer, but the lands that became Europe, Africa, Asia, and the Americas were all connected by warm shallow seas. *Sepioteuthis* likely had a wide distribution, through the remnants of the Tethys Sea, and its extensive range has persisted to the present. Other genera evolved more recently, after the development of the Atlantic and Pacific Oceans, and the closure of the Mediterranean Sea. These younger genera are restricted to smaller areas, because warm shallow connections between these locations no longer exist.

FUZZY NAUTILUS

Allonautilus scrobiculatus

FAMILY	Nautilidae
OTHER NAMES	Crusty nautilus
TYPICAL HABITAT	Outer slopes of coral reefs to 2,300 ft (700 m) depth
SIZE	To about 8 in (20 cm) diameter
FEEDING HABITS	Scavenger/opportunistic predator
KEY BEHAVIORS	Vertical migration

THIS INCREDIBLY RARE NAUTILUS, WHOSE shell has hardly changed over hundreds of millions of years, is only known from the waters around Papua New Guinea. Nautiluses are the only group of living cephalopods to have an external shell. The extensive hairy outer shell layer (the periostracum) of the fuzzy nautilus gives it a highly distinctive appearance compared with all other nautilus species and also gives this species its common name. It has been recognized as a unique species since the 1700s based on drift shells that have washed ashore. Nonetheless, the fuzzy nautilus was not seen alive until 1984 and an absence of reports of the species over the next 30 years led scientists to suspect that this "living fossil" might have finally gone extinct. It was rediscovered in 2015 off Ndrova Island, Papua New Guinea.

A LIFE ON THE REEF'S EDGE

Nautiluses live on the outer slopes of coral reefs. They spend much of their time in deep waters, rising to shallower depths to feed. By staying in darker deeper waters whenever possible, they may avoid large predators or they may benefit from a slower metabolism that reduces their resource requirements. The shells of nautiluses are divided internally into chambers and the animals control their buoyancy by adjusting the gas content of these chambers. Despite their large eyes, nautiluses use their sense of smell to detect their prey, which they handle with their many tentacles.

A DISTINCTIVE SHELL

The fuzzy nautilus is easily distinguished from other nautiluses by its yellowy brown proteinaceous outer layer of shell. The shell is also highly coiled with a deep umbilicus—the depression at the center of the shell whorls.

COMMON BLANKET OCTOPUS

Tremoctopus violaceus

THE BLANKET OCTOPUS IS ONE OF a small group of species known as argonautoids (superfamily Argonautoidea). Perhaps most famous for their detachable mating organ, and their extreme sexual size dimorphism, as their group name suggests, argonautoids are "sailors"—exhibiting a pelagic lifestyle in contrast to their closest relatives, the benthic octopods. Like most other argonautoids, the blanket octopus has a hydrostatic organ deep within its viscera to give it buoyancy. The dorsal arms of females produce secretions that harden to a calcareous rod that the female carries and to which she attaches her eggs: a neat solution to brooding away from the seafloor.

DEFENSE

Juvenile blanket octopods often carry fragments of Portuguese man o' war siphonophores, complete with dangerous stinging cells, presumably as defense. It is not known how the *Tremoctopus* protects itself and it is possible that the difficulty of manipulating the stingers with the larger suckers found in a growing octopus promoted dwarfism in males. Larger female size is dictated by the need to produce and brood eggs, but adult females have developed their own defensive strategy. The dorsal arms are longer, and between them is an extensive web patterned with ocelli. When threatened, the female drops part of these arms, as a lizard drops its tail, leaving a small floating piece of web with markings that may even mimic the female's egg masses, allowing her to escape.

FAMILY	Tremoctopodidae
OTHER NAMES	Violet blanket octopus
TYPICAL HABITAT	Epipelagic zone (where light penetrates)
SIZE	Females to 16 in (40 cm) mantle length, 5 ft (1.5 m) total length, males to 13/16 in (2 cm) mantle length
FEEDING HABITS	Eats small nekton and plankton including fishes and pteropods (swimming snails)
KEY BEHAVIORS	Arm dropping (autotomy), including autotomy of male hectocotylus

WIDESPREAD WEB
A female common blanket octopus is seen here in surface waters with her arms and web widely spread. The blanket is spread between the dorsal-most arms and the second and third arms. Note the ocelli, the small circles that run in lines from the second arm pair, with each ocellus decreasing in size relative to its neighbor.

VAMPIRE

Vampyroteuthis infernalis

FAMILY	Vampyroteuthidae
OTHER NAMES	Vampire squid
TYPICAL HABITAT	Temperate and tropical oceans from 2,000 ft (600 m) to 4,000 ft (1,200 m) depth in the oxygen minimum zone
SIZE	Mantle length to 5 in (13 cm), total length to 12 in (30 cm)
FEEDING HABITS	Detritivore, perhaps including other food types
KEY BEHAVIORS	Bioluminescence, pineapple posture

THE VAMPIRE IS A "LIVING FOSSIL": the sole survivor of a group of animals that diversified in the Jurassic and Cretaceous periods. It is most closely related to octopods. It has ten arms, but two are highly modified giving it the appearance of an eight-armed creature. The vampire trails its two long, sticky, retractile filaments through the ocean collecting food particles. It is the only known cephalopod that feeds on "marine snow"—the aggregated debris of decaying plankton, crustacean molts, and fecal pellets, yet it also consumes small live crustacean prey such as copepods, which inhabit and feed on the aggregates. Although juveniles may use jet propulsion, adults flap their fins to swim. When startled, vampires can turn their deep web inside out over the rest of their body in a "pineapple posture." Vampires have the largest eyes relative to body size of any animal.

BIOLUMINESCENCE

Vampires have interesting mechanisms of defense at their disposal. They have light-producing organs at the bases of their fins and on the tips of their arms that can glow, flash, or vary in light intensity such that the light appears to pulsate. Vampires produce luminescent displays in response to mechanical stimuli, but a captive male has been seen to respond to a captive female's display, suggesting a role in communication as well as defense. The arm tips can also produce bioluminescent clouds of glowing particles bound in a viscous fluid—a behavior unknown among other cephalopods.

AT ARM'S LENGTH

The vampire is deeply pigmented, with a pair of fins for swimming, and a deep web between its eight unmodified arms. Each arm is adorned with suckers and cirri, which help to pass food from the retractile filaments to the mouth.

TWO-TONED PYGMY IDIOSEPIID

Idiosepius pygmaeus

FAMILY	Idiosepiidae
OTHER NAMES	Thumbnail squid, two-toned pygmy squid
TYPICAL HABITAT	Shallow coastal seagrass beds
SIZE	Males to ½ in (12 mm) mantle length; females to 11/16 in (17 mm) mantle length
FEEDING HABITS	Predator on small planktonic crustaceans
KEY BEHAVIORS	Counter shading; adhesion

THESE TINY CREATURES ARE FOUND in shallow coastal seas, particularly among seagrass beds, from northern Australia, through the Indo-West Pacific to as far north as the Philippines and as far west as Thailand. They are sexually dimorphic, with females larger than males, but even the females are less than 13/16 in (2 cm) long. They lay tiny eggs, attached in batches to shells or rocky substrate. These hatch in just 1 or 2 weeks into tiny planktonic hatchlings with mantles 1/24 in (1 mm) long. Growth is rapid: they live for 10 weeks, in which time they have reached adulthood and reproduced, possibly laying multiple batches of eggs. Despite their tiny size, they are active predators, taking glass shrimps and possum shrimps from the plankton.

A STICKY BACK

A peculiar feature unique to idiosepiids is the glue gland on the dorsal surface of their mantle near the fins. Some other cephalopods secrete adhesive substances, but none has perfected it quite the way idiosepiids have. They can literally glue themselves to a blade of seagrass or alga. This can provide protection from predation through camouflage, or simply help them to conserve energy through not having to swim against any current. The glue gland, obvious by its rougher surface, is composed of two types of glandular cells that secrete the carbohydrates and proteins that form the glue.

SMALL AND HIDDEN
The chromatophores on this tiny cephalopod give an idea of how small it actually is, even in the absence of another feature for scale. Attached to a seagrass blade, it is barely visible.

CARIBBEAN REEF OCTOPUS

Octopus briareus

FAMILY	Octopodidae
OTHER NAMES	Briar octopus
TYPICAL HABITAT	Coral reef and associated seagrass beds
SIZE	Mantle length to 4¾ in (12 cm); total length to around 24 in (60 cm)
FEEDING HABITS	Nocturnal predator of crustaceans
KEY BEHAVIORS	Pounce, deimatic threat/startle displays

THIS REEF-DWELLING OCTOPUS INHABITS shallow tropical waters of the Caribbean and western Atlantic, its distribution extending from southern Florida and the Bahamas to northern South America. Caribbean reef octopuses are nocturnal ambush predators, easily recognizable by the greenish-blue tinge on their extensive web; they feed on crustaceans such as crabs and lobsters. These crustaceans are far from defenseless. Caribbean spiny lobsters stridulate (produce sound by moving specialized body parts) when attacked by Caribbean reef octopuses and lobsters that stridulate escape from octopus attacks more frequently than those that remain silent. Like other large-egged species, the hatchlings of Caribbean reef octopuses are not planktonic, but immediately begin life on the seafloor like miniature adults. Because Caribbean reef octopuses have no planktonic juvenile stage, this is one of the easier species of cephalopod to culture in captivity.

ROLE OF DENS

During the day, Caribbean reef octopuses can be found hidden in their dens, which tend to be small natural crevices in the reef structure that provide shelter. They are a solitary species and defend their dens from their fellow octopuses. Females also lay their large eggs in clusters in their dens, and often block the entrance with shells from dead clams and pieces of coral. Caribbean reef octopuses not only prey upon Caribbean spiny lobsters, but also compete with them for dens. When shelter sites are limited, octopuses are dominant.

TRAPPING PREY
The posture seen here is typical of a "pounce" attack. In this hunting strategy, the octopus accelerates toward the prey using jet propulsion, and then drops, parachute-like, with its arms widely spread, trapping the prey underneath its extensive web. The octopus then reaches under its web with its long arms to manipulate the prey toward its beak.

FLAMBOYANT CUTTLEFISH

Metasepia pfefferi

FLAMBOYANT CUTTLEFISHES ARE FOUND off the more northerly coasts of Australia and off Papua New Guinea and other islands of the Indo-West Pacific. They live on sandy and muddy bottoms in shallow, coastal waters where they are typically well camouflaged and look unremarkable. Their cuttlebone is smaller than in typical cuttlefishes, and this might contribute to their reduced buoyancy and hence their ability to "walk" on the sea floor. This unusual locomotion involves their most ventral pair of arms. They may raise themselves up on these arms to give themselves added height, or they may curve these arms under their body and step them one over the other.

MASTERS OF COLOR CHANGE

Flamboyant cuttlefishes are capable of producing "traveling wave" or "passing cloud" displays. These are bands of color that move continuously across their skin. Although other cephalopods can produce these displays, laboratory tank experiments suggest that flamboyant cuttlefishes are able to keep the display up longer and use it during more diverse activities, including swimming, walking, and mating. When disturbed, flamboyant cuttlefishes exhibit bright warning colors. Although warning colors are often associated with toxicity, such that a predator learns to recognize the warning, the tissues of the flamboyant cuttlefishes have been found to contain only minute traces of tetrodotoxin, certainly not enough to stun or kill its predators.

FAMILY	Sepiidae
OTHER NAMES	Pfeffer's flamboyant cuttlefish
TYPICAL HABITAT	On sandy and muddy bottoms to 330 ft (100 m) depth
SIZE	2⅜ to 3⅛ in (6 to 8 cm) mantle length
FEEDING HABITS	Actively prey on fish and crustaceans
KEY BEHAVIORS	Unique ambulatory movements and rapid color change; warning coloration

WARNING COLORS

A flamboyant cuttlefish holds its ground when startled or threatened and flashes its warning coloration to deter would-be predators. The animal raises papillae and displays alternating dark and white patches on the dorsal surface. When danger is averted it returns to its normal camouflage.

ANALOGOUS BOBTAIL

Sepiola affinis

FAMILY	Sepiolidae
OTHER NAMES	Analogous bobtail squid
TYPICAL HABITAT	Sandy bottoms of continental shelves from 50–500 ft (15–150 m) depth, but mostly found between 50 and 100 ft (15 and 30 m)
SIZE	Mantle length to 1 in (2.5 cm)
FEEDING HABITS	Small crustaceans
KEY BEHAVIORS	Bioluminescence; burying

BOBTAILS SEEM TO HAVE UNDERGONE rapid evolutionary radiation in the Mediterranean Sea and this is one of several species that are endemic (found nowhere else). *Sepiola affinis* has all the classic attributes of a bobtail—a round mantle, rounded fins, and a rudimentary gladius. It lives in shallow coastal waters on sandy and muddy bottoms. It has two light organs (photophores) inside the mantle on the viscera, which house the symbiotic luminescent bacteria *Vibrio logei* and *Vibrio fischeri*. Typically, this bobtail spends the day buried in soft sediments, emerging at night to feed. At night, the photophores illuminate the underside of the bobtail, not to make it more visible, but to camouflage its shadow from moonlight and starlight penetrating surface waters.

BODY PATTERNS

For a nocturnal animal, *Sepiola affinis* exhibits a broad range of body postures, colors, and patterns. These are particularly noticeable during feeding when this small bobtail undergoes rapid color changes, particularly as it strikes out with its tentacles. It can range from very pale to completely dark, have white edges to otherwise dark fins, a dark head bar, or dark spots on the mantle rim. It also exhibits complex dynamic displays such as "passing cloud" whereby dark bands of color travel along the length of the body.

EXTENSIVE CAMOUFLAGE

The rich color patterns displayed by this bobtail are possible because of the dense chromatophores that cover its whole body. By day, using a combination of arm and fin movements and jets of water from its funnel, the analogous bobtail can rapidly bury itself in sediment so that just its eyes are uncovered, and no further camouflage is required.

OPALESCENT INSHORE SQUID

Doryteuthis opalescens

FAMILY	Loliginidae
OTHER NAMES	Californian market squid
TYPICAL HABITAT	Shallow coastal waters
SIZE	Mantle length to 7½ in (19 cm); total length to 12 in (30 cm)
FEEDING HABITS	Opportunist. Takes a wide range of prey including crustaceans, fishes, and other cephalopods
KEY BEHAVIORS	Mass spawning; male dominance behavior

THIS SMALLISH SQUID INHABITS THE shelf waters of western North America, where it can be massively abundant. They are prey to numerous fishes, seabirds, and marine mammals, and therefore play an important role in the food web. Their eggs, which are laid on the seafloor, take about a month to hatch, and the squids take a further 6 to 9 months to mature into adults. Mature opalescent squids aggregate on spawning grounds from California to British Columbia, the time of year being dependent on latitude and temperature. In some areas, spawning aggregations can involve millions of animals. In other places, such as Monterey Bay, the aggregations are smaller but may still involve thousands of squids in the water column above smaller spawning groups closer to the seafloor. Males perform postural and color displays to male competitors for access to female mates, which they guard carefully while temporarily paired with them during egg laying.

EL NIÑO

The current system in which these squids live is affected by the El Niño climate cycle, which disrupts the upwelling of cold nutrient-rich water that supports high primary production. During El Niño events, ocean productivity is depressed and squids grow more slowly to smaller maximum sizes and are less abundant. Conversely, during the reverse La Niña events, squids are larger and more plentiful.

CAUGHT RED-HANDED

When mating, a male opalescent inshore squid grasps the female firmly and holds her nearly vertical both while mating and while she subsequently lays her eggs on the seafloor. Males are easily distinguished when spawning as their arms redden to a dark hue in comparison with the more uniform color of the female.

EYE-FLASH SQUID

Abralia veranyi

ORIGINALLY DESCRIBED FROM THE Mediterranean Sea, this small species of oceanic squid is found on the continental slope of both the western and eastern Atlantic and above slope habitats offshore, for example, around the Bahamas and some seamounts. Living in mid-water, the eye-flash squid migrates vertically through the water column, spending the day at deeper darker depths, perhaps avoiding predators such as tunas and dolphins. At night it rises to shallower depths where it likely feeds on small crustaceans.

DIMMING THE LIGHTS

The name eye-flash squid derives from the photophores around its eyes. It has five of these large light-emitting organs on the ventral (under) side of the eye; one of these is extremely large and produces bright flashes. However, the underside of the mantle is also completely covered in photophores, with up to 500 or more of these specialized organs. *Abralia veranyi* uses these photophores to break up its silhouette against lighter surface waters when viewed from the darker water below—a mechanism known as counterillumination. Remarkably, these squids can adjust the wavelength of light that their photophores emit, shifting the color of their light emissions to maximize camouflage.

FAMILY	Enoploteuthidae
OTHER NAMES	Verany's enope squid
TYPICAL HABITAT	Surface waters to 2,620 ft (800 m) depth above slopes and seamounts
SIZE	Mantle length to 2 in (5 cm)
FEEDING HABITS	Hunt at night in surface waters
KEY BEHAVIORS	Bioluminescence

BRIGHT EYES

The very large posterior eye photophore is prominent in this photograph of *Abralia*. It is this photophore that flashes and gives the squid its name. The natural posture of *Abralia* is unknown and the posture seen here may be a defensive response to the bright lights and equipment; however, it should also be noted that this is a hydrodynamically stable posture for sinking.

PECULIAR LIFESTYLES

REPRODUCTION

AN ENDURING MISCONCEPTION ABOUT cephalopods is that they all spawn once and then immediately die. This pattern may be true for a few species that are convenient for study and has been publicized in popular media. Part of the reason for this publicity may be that people tend to think it somehow wasteful that these intelligent invertebrates "live fast and die young." However, as we learn more about diverse cephalopods, we find much variation on this theme.

THE SPAWNING CYCLE

Reproducing once and then dying is the extreme situation of a life-history pattern in which the sexual organs mature only once during the life cycle. The opposite of this is when sexual organs mature and regress in cycles; nautilids are the only cephalopods known to have such cycles. However, not all cephalopods reproduce in a single "big bang." More and more species are being shown to spawn differently. Although their reproductive organs do not undergo cycles of regression and regrowth, spawning is prolonged either by slow continuous release of eggs (as in the family Enoploteuthidae and cirrate octopods) or by periodic maturation and release of many eggs (in some oceanic squids). Among other things, whether a species is a big-bang or a batch spawner is important in management of commercial fisheries for cephalopods.

MATING

As with many aspects of cephalopod biology and behavior, most of the detailed observations of mating involve common coastal species in the families Octopodidae, Loliginidae, and Sepiidae. In general, cephalopod mating does not involve long-term association of mated pairs. A possible exception to this is a large but fairly uncommon oceanic species, the diamondback squid *Thysanoteuthis rhombus*, the only species in the family Thysanoteuthidae. Rather than schooling, these squids are often found in a pair consisting of a male and a female. If one of the pair is caught by squid fishermen, the other member is likely to remain in the area until it is also caught. Most other species that have been observed, however, pair up only for a brief period. For most octopods, this period may comprise just the time (a few minutes to several hours) required for recognition, courtship, and spermatophore transfer. For inshore squids, it may include guarding of the female by the male until eggs are deposited.

Mate guarding is likely a defense against sperm competition. Such competition occurs when the sperm in a mated female is displaced by sperm from another male in a subsequent mating before the eggs are actually fertilized. Because the sperm is transferred in masses packaged in spermatophores, some time is required for the spermatozoa actually to reach an egg and fertilize it. The processes by which spermatozoa move to the eggs from the sites of the embedded, discharged spermatophores in various species is not well understood. It is rapid in some species, while in others females are known to store sperm for prolonged periods of up to several months.

Mating has been studied in detail for a few coastal species of octopods, cuttlefishes, bobtails, and inshore squids. Most octopods are solitary creatures and seem to mate without forming any sort of pairing behavior; that is, they encounter one another, mate, and depart. There are a few field anecdotes of a male and female octopus residing next to each other and mating, sometimes for two to three days in succession, but this is rare. Conversely, many species of cuttlefishes and inshore squids form temporary pairs (lasting minutes or hours, but not days) and a few participate in communal spawning activities. In the spawning aggregations, large males compete for females. Competition consists of shoving matches

Opposite top A female of the common octopus (*Octopus vulgaris*) attaches a mass of eggs to a hard structure and then cares for them, while her condition deteriorates, until they hatch. She dies soon afterward.

Opposite bottom A mating pair of octopuses. The male (left) is reaching out to the female (right) with his hectocotylus, which is inserted into her mantle opening to deposit spermatophores in her oviducal gland.

Left A male Caribbean reef squid (*Sepioteuthis sepioidea*) transferring spermatophores to a female.

Opposite Head-to-head mating by cuttlefish *Sepia apama*.

as well as characteristic displays of chromatophore patterns to advertise dominance and to intimidate rival males (see Chapter 4). The successful male will mate with the female and then guard her temporarily against mating by other large males.

In many of these species males differ in size and behavior at spawning time. Rather than fighting with the big males for mating privileges with females, the small "sneaker" males use different tactics, one of which is "female mimicry" (Chapter 4). Overall, cephalopods have a mostly promiscuous mating strategy. That is, both males and females have multiple mates, although females in particular are choosy about their mates. One result of the multiple mating tactic is that spermatophores may be implanted on the females in different places by dominant males and sneaker males. If a dominant male loses his contest and is replaced by another dominant male, mating will occur and the female can have another source of sperm. DNA fingerprinting studies show that a female can use multiple sperm sources, even though they may be implanted or stored on different locations in her anatomy. This phenomenon is known as "cryptic female choice" because it occurs after mating and when no male is around (so it is cryptic to the male). DNA studies of the egg masses of inshore squids and cuttlefishes have demonstrated multiple paternities among the eggs in a single egg strand or egg clump.

Regardless of species or the behaviors leading to copulation, the male transfers his spermatophores to the female. The male

incirrate octopod inserts the tip of his hectocotylus into the female's mantle cavity. Spermatophores travel down the spermatophore groove along the side of the arm and are implanted in the oviducal glands of the female. There is speculation that the male incirrate octopod uses the modified tip of the hectocotylus to remove any discharged spermatophores from previous matings of the female, but the evidence for this form of sperm competition is not yet clear. Nothing is known about cirrate mating, except that the males have neither hectocotylization nor elongate penises and that the spermatophores are highly modified, tiny sperm packets.

Sexual size difference becomes extreme in some of the pelagic incirrates in which the males are dwarfs, but have comparatively large elaborate hectocotyli. An extreme case of hectocotylus elaboration is the *Argonauta* spp., in which the large hectocotylus carrying a single large spermatophore breaks off in the female, retaining the ability to crawl about. When this vermiform wiggler was first discovered it was described as a new genus of worm, *Hectocotylus*.

Some oceanic squids lacking hectocotylization have a greatly elongated terminal segment ("penis") on the male reproductive tract, which serves the same function as the hectocotylus. Recent observations from a robot submersible of large deep-sea scaled squid showed that such coupling can be prolonged, although it has been speculated to be very quick in other species, including giant squid.

EGGS & SPERM

The eggs of all living cephalopods, and presumably extinct ones as well because of the similarities between living nautiloids and neocoleoids, include a large yolk enclosed in a primary membrane (chorion). The oocytes develop from the epithelium of the single ovary, enveloped in a highly vascularized, folded follicular cell layer. The size of the eggs and the amount of yolk vary among neocoleoid species, in some cases among closely related species (such as *Octopus* spp.). Once the egg is mature and the female ready to lay, the follicle ruptures and the egg is released into the open area of the ovary. Eggs are "wrapped" in a variety of ways for protection; eggs of incirrate octopods are cared for by the female, and in a few deep-water squids such as gonatids and bathyteuthids the females tow their egg masses around with them. The eggs of all other cephalopods are left by the mother in nature and have to develop and hatch on their own without any parental care; they are exposed to all surrounding seawater conditions as well as to potential predators.

Sperm is produced in the single testis, which occupies the same location in the male as the ovary does in the female. After the spermatozoa have fully differentiated and been released by the testis, they pass through a series of tube-like ducts and multiple organs that build the spermatophore. Sperm are first packaged into a coiled mass with an "ejaculatory apparatus" that will eventually release them from the spermatophore after transfer to the female, and then the entire package is enclosed by a non-cellular tunic. The completed spermatophores are stored in a sac near the end of the male reproductive tract until the male is ready to release them through the terminal organ (penis), which may be a simple duct or a long tube depending on whether the species has a hectocotylus. Curiously, although the male reproductive tract is generally single in cephalopods, the histioteuthid genus *Stigmatoteuthis* has paired tracts similar to the paired female tracts found in many groups.

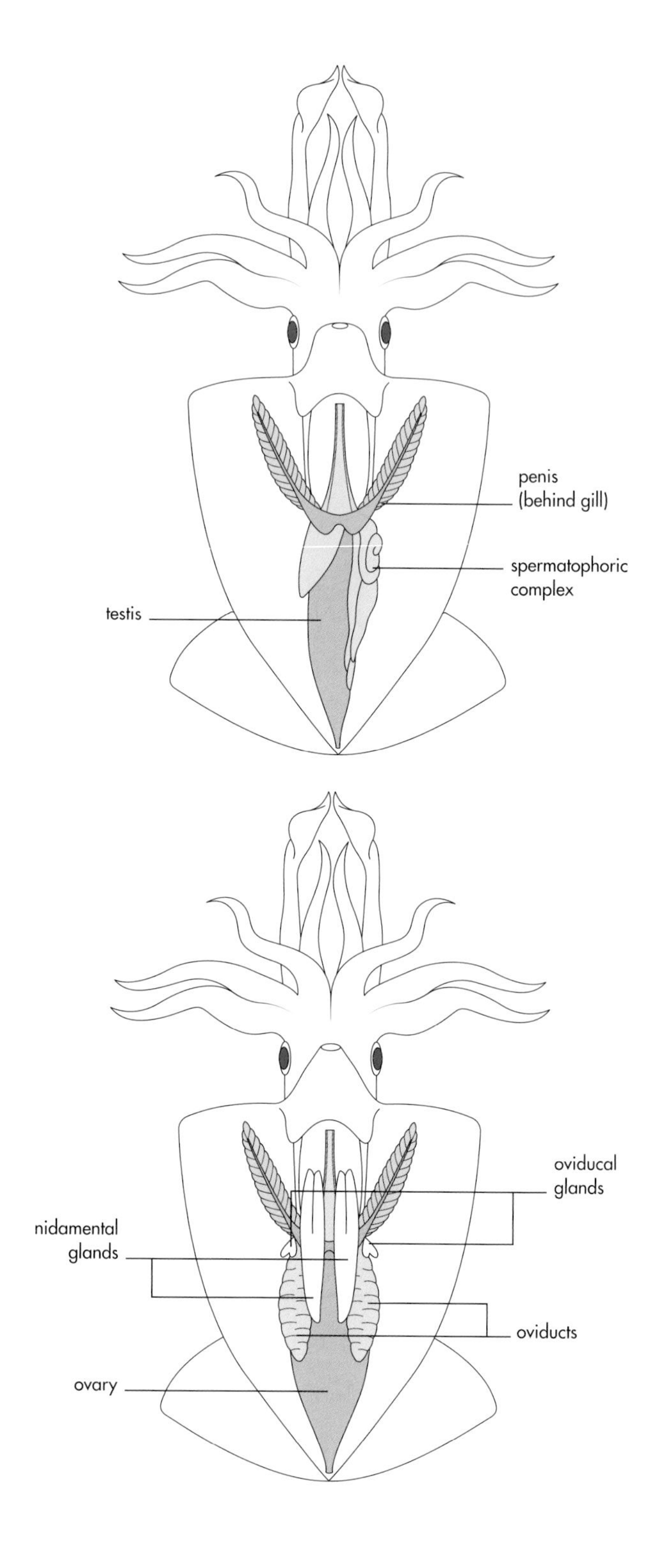

Above The reproductive systems of an oceanic squid. Male (top), female (bottom).

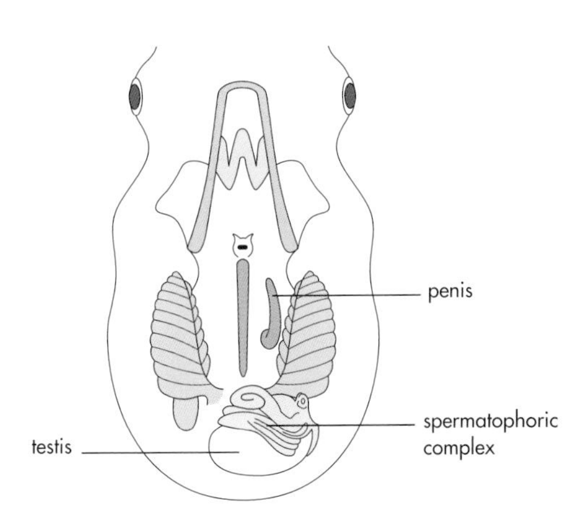

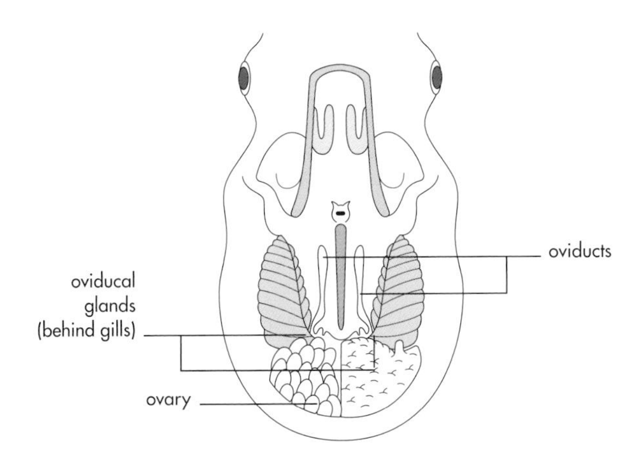

Above Incirrate octopod reproductive systems. Male (top), female (bottom).

Above right and right The male reproductive system of oceanic squid genus *Stigmatoteuthis* (top) is unusual because it has two sets of ducts, spermatophore producing organs, and penises (bottom).

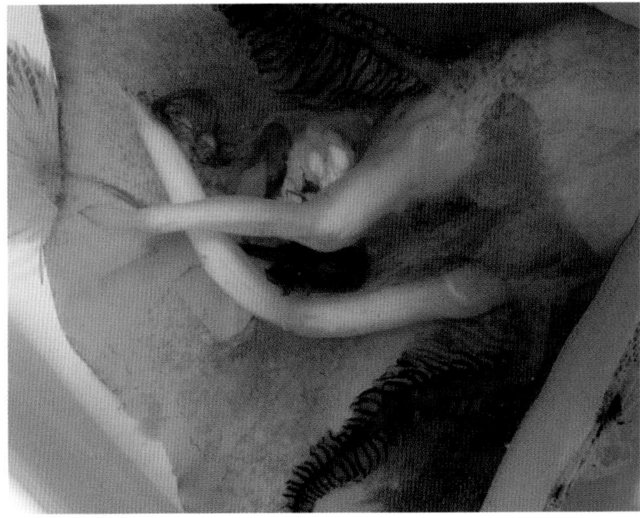

FERTILIZATION

Insemination varies among species of decapods. In inshore squids, the male holds the female, while batches of spermatophores are grasped by the male using the tip of the hectocotylus, and then jammed against the female in particular locations, such as around the mouth or the inside of the mantle cavity near the opening of the oviduct, where the spermatophore discharges and the sperm mass is implanted. Spermatophore implantation is similar in sepiids. Female bobtails have a pocket inside the mantle into which spermatophores are implanted. Mating has seldom been observed among the diverse oceanic squids. Sites where spermatophores are implanted on the females indicate some strange patterns. Depending on the species, spermatophores are implanted inside the mantle, around the mouth, on the arms, at the nape of the neck, and even on the outside of the mantle. How the eggs are exposed to sperm by the female is unknown.

The group in which fertilization makes most sense is the incirrate octopods. Here, the spermatophores are implanted in the oviducal glands, so the eggs encounter sperm masses early in the journey; there are no extra-chorionic coatings that could prevent penetration of a sperm into the egg. This pattern has not been clearly demonstrated in cirrate octopods. Although the oviducal gland in cirrates secretes a coating on the egg before it moves into the distal oviduct, presumably sperm packets are taken into the oviduct or oviducal gland as in the incirrate octopods and fertilization can occur in either the oviducts or in the oviducal gland before the coating.

Decapods generally have both oviducal and nidamental glands and some have accessory nidamental glands. Their eggs are coated by thick secretions, first by the oviducal gland and then the nidamental gland. The sperm masses are implanted at various locations distant from the oviducts, so the sperm must either be transferred up the oviduct beyond the oviducal gland or it must be able to swim through the viscous secretions of the oviducal gland and perhaps the nidamental gland. The only evidence of how this happens comes from some incidental observations of egg laying by species of Ommastrephidae. In this family, spermatophores are implanted either around the female's mouth or inside the mantle cavity. Large gelatinous egg masses are formed when the female mixes seawater with nidamental jelly as the mixture of eggs and jelly is extruded by the female through her funnel. Broken spermatophores have been reported mixed with the eggs in the gelatinous egg masses. It is known from in vitro fertilization experiments that these eggs are not fully hydrated and competent for fertilization until after they have been exposed to oviducal gland secretions for several minutes.

So it seems that the sequence of events leading to fertilization in this family is:

1) batches of eggs accumulate in the oviduct

2) as they pass through the oviducal gland they are coated with oviducal gland secretions, allowing them to hydrate and preparing them for fertilization

3) they are then mixed with nidamental gland jelly in the mantle cavity and this is mixed with sperm either in the mantle cavity or near the mouth as the egg mass is expanded with water.

A problem with this scenario is that, based largely on observations of inshore-squid egg masses (which admittedly are quite different from those of ommastrephids), a function of the nidamental jelly is to prevent microbial infestation by inhibiting ciliary and flagellar motion. It is not yet clear how well squid sperm swim in this jelly. Even if the scenario described above is correct for the family Ommastrephidae, it must be remembered that ommastrephids and their egg masses are very different from other oceanic squids, not to mention the more distantly related inshore squids and cuttlefishes; what works for ommastrephids may not hold true for other decapods.

Opposite Female bobtails, like this *Sepiola atlantica*, have a pocket inside the mantle into which spermatophores are placed by the male.

Below An egg of a cirrate (finned) octopod, deposited individually on a deep-sea coral without further maternal care. This egg has burst its tough outer coating because it has swollen during development of the embryo.

Below right The egg mass of an incirrate octopod, attached to the substrate and cared for by the mother, seen here below her eggs.

EGG LAYING

Nautilids lay very large eggs contained within a hard coating. These eggs are deposited in crevices. Although the eggs of neocoleoids are basically similar to each other, the structure and appearance of the egg masses differ greatly among the major groups. This is, in some ways, related to the differences in the female reproductive systems. Incirrate octopods with oviducal but no nidamental glands lay eggs without complex outer coatings, but eggs of cirrate octopods are cemented individually to benthic structures with a thick coating secreted by the oviducal gland. The chorions of incirrate eggs are drawn out at one end into a thin string that the female weaves together with the strings from other eggs to form strands of eggs like bunches of grapes. The females of most benthic incirrate species attach their egg strands to benthic substrate, often inside their dens, and then aerate and protect them until hatching. The males do not participate in this parental care.

Above Instead of attaching their eggs to hard substrate, females of the southern blue-ringed octopus carry them in their arms.

In a few benthic incirrate species the females carry their eggs around with them. In all pelagic incirrates the females carry their egg masses. Several species hold their eggs within the arm webs; in Tremoctopodidae the eggs are attached to calcareous rod-like structures that are held in the arms of the female. Families Ocythoidae and Vitreledonellidae actually appear to have evolved forms of ovoviviparous development. The former retain the fertilized eggs in the oviducts, whereas the latter supposedly brood their eggs within the mantle cavity through to hatching.

Among the decapods, inshore squids, bobtails, and cuttlefishes lay benthic eggs. Inshore squids package their fertilized eggs into multilayered, gelatinous, finger-like strands that are attached to the benthic substrate in clusters. Cuttlefish eggs are larger than those of inshore squids and are not embedded in gelatinous masses. The eggs of cuttlefishes are laid individually, each packaged in several layers of tough protective coatings, but most species lay many eggs together in clusters. Some cuttlefish species inject ink inside the egg coating, thus darkening the appearance to the egg capsule; others cover eggs with a thin layer of sand to camouflage them; yet others lay eggs under flat rocks to hide them from predators. The multiple gelatinous layers of inshore-squid egg strands and the multiple hardened layers of cuttlefish eggs are secreted by the nidamental glands in addition to the oviducal glands of the females, an important difference from octopods.

The eggs and egg masses are known for only a few oegopsid families. The floating egg masses of *Thysanoteuthis* are frequently encountered by divers. They appear somewhat similar to those of the family Ommastrephidae, which are diaphanous spheres of jelly, up to 3 ft (1 m) in diameter. However, the overall shape of a *Thysanoteuthis* egg mass is more elongate (more "sausage-shaped") and the eggs are organized in a long strand that spirals tightly around the subsurface layer of the mass. Family Enoploteuthidae, which lack functional nidamental glands, lay individual planktonic eggs that are not coated in jelly.

The recent discovery of maternal care by the families Gonatidae and Bathyteuthidae illustrates how little confidence can be placed in generalization of patterns observed for a few families to broader cephalopod groups in general. Because female incirrates care for their eggs but several decapod families are known not to do so, it had been assumed that maternal care was a fundamental difference between decapods and octopods. It was totally unexpected that in some oceanic squids the female would lay an egg mass and then tow it around in her arms. Given the diversity of oceanic-squid families, it will be interesting to see how many other biological surprises await us.

Left The egg mass of an oceanic squid, the diamondback squid *Thysanoteuthis rhombus*, floating near the surface. Squid predators (whales) are seen in the distance.

Below In some families of oceanic squid, like this gonatid, the female tows the egg mass in deep water until the eggs hatch.

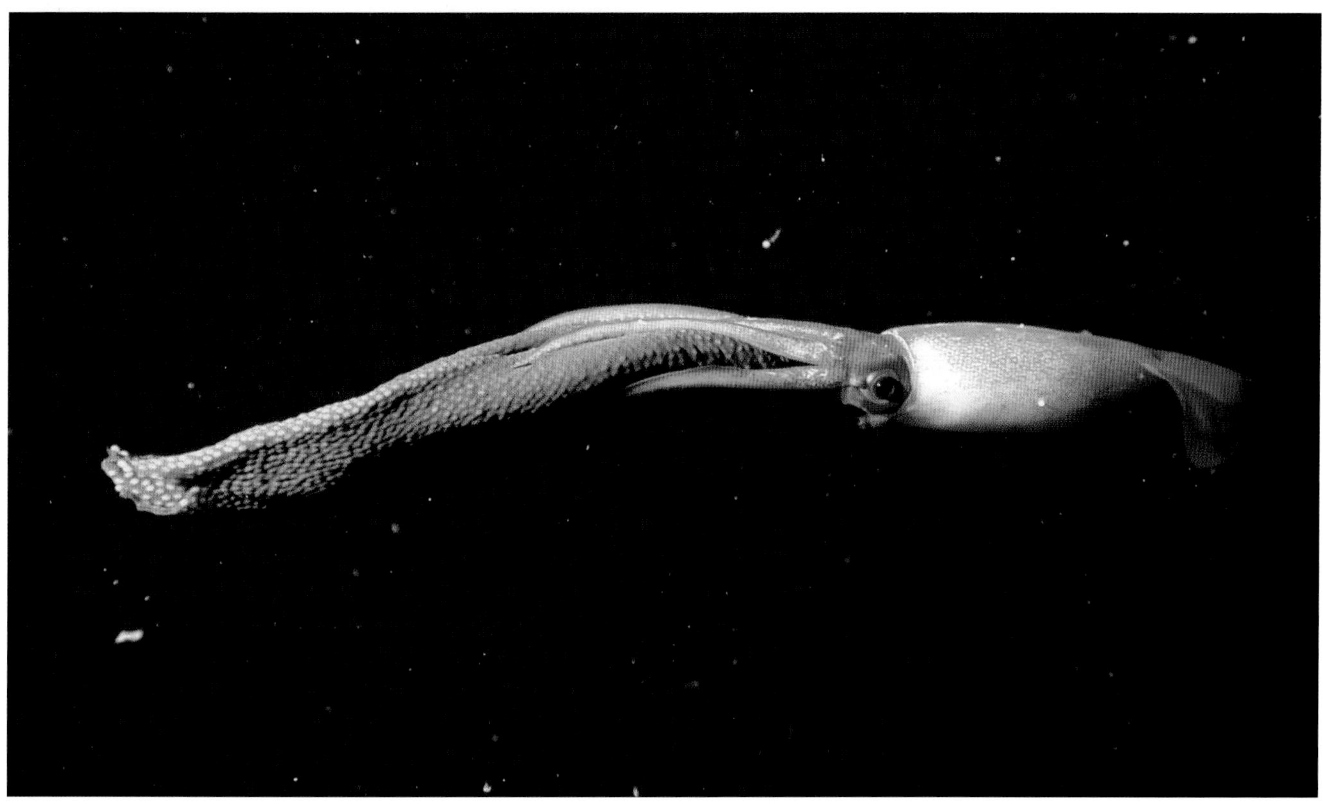

DEVELOPMENT

DESPITE DIFFERENCES IN EGG SIZE among cephalopods, all can be characterized as large and yolky compared with other molluscs. A result of the large yolk mass is that the embryo develops initially as a disk at one end of the yolk. Whereas other molluscan eggs undergo complete spiral cleavage similar to annelids and arthropods, cephalopod cleavage is very different from its beginning.

NOT VERY MOLLUSCAN

Most molluscs develop through, and many hatch as, a primitive ciliated larval stage, followed by a second larval stage. No remnant of either stage is found in cephalopods, whose embryos develop without any special larval organs. Therefore, cephalopods undergo direct development.

Cleavage in cephalopod eggs is bilaterally symmetrical and initially incomplete; the first few cell divisions form furrows in the surface of the animal pole of the egg but do not result in cells that are completely surrounded by cell membranes. Subsequent cell division forms complete and separate cells in a single layer across one end of the egg. The bilaterally symmetrical cleavage of cephalopods should not be confused with the radial cleavage characteristic of animals such as echinoderms and chordates. In most animals, cleavage results in a ball of cells known as a blastula. In cephalopods, the layer of cells becomes a densely packed, cap-like disk covering the end of the egg, very different from a blastula.

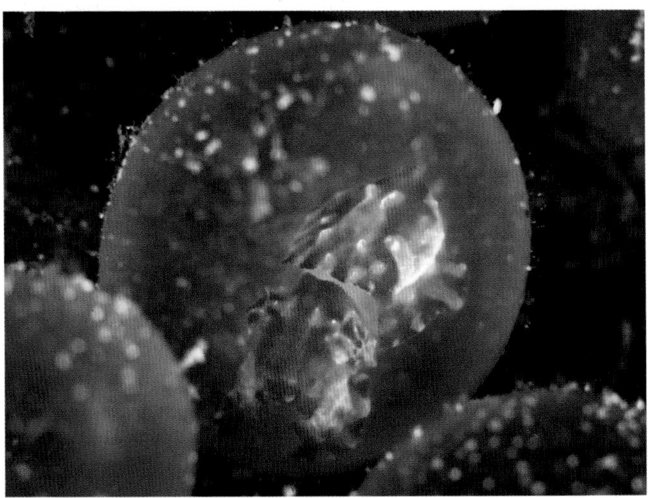

HOW AN EMBRYO BECOMES A CEPHALOPOD

The head and its organs develop in the upper part of the disk, while the arm crown develops around the periphery; organs such as the gills and funnel develop in the lower part of the disk; and the shell gland and sac, mantle, and fins arise in the center. The ring of arm development is separated bilaterally by the developing mouth at the embryological anterior and the anus at the posterior. Each arm begins as a pair of bumps. Another curious and unexplained phenomenon is that the dorsal-arm precursors are broadly separated from those of the rest of the arms. As the arms elongate, bumps forming on their oral surfaces will become the suckers in neocoleoids and the adhesive ridges of nautilids. Tentacles of decapods are initially indistinguishable from the other arms. There is no development of the missing pair of arms in octopods, whereas the tentacle precursors in family Ommastrephidae shift together in position to form the fused appendage characteristic of ommastrephid hatchlings and paralarvae. The mantle of neocoleoids begins as a ring of tissue, in the center of which the shell gland and sac develop. The outer edge of the mantle then grows outward as it elongates to cover the developing gills and guts while the inner edge closes over the shell sac. The fins of those species that have them develop as the embryo elongates and becomes more cephalopod-like. Late in neocoleoid development, the chromatophores and ink sac develop. The chromatophores are active before hatching.

As the embryo grows, it consumes the yolk, which eventually forms two separate but connected masses, the external yolk sac and the bilobed internal yolk sac. Yolk is then gradually transferred from the external to the internal yolk sac. By the time of hatching the yolk is entirely internal. Mysteriously, the embryo reverses its orientation within the egg not once but twice, returning to its original orientation before hatching.

Duration of embryonic development depends on two factors: egg size and temperature. Larger eggs produce larger embryos that take longer to develop, and warmer waters can substantially shorten development times. Small-egged tropical species can develop fully in a few days whereas large-egged polar and deep-sea species may take over a year. Most temperate-climate nearshore cephalopods develop in one to three months. A deep-sea octopod was recently shown to take at least four years for embryonic development. Even though nautilids develop in fairly warm water, their very large eggs require eight to twelve months to develop.

Opposite top Close-up of the egg mass of an incirrate octopod. The eggs are connected in multiple strings and the developing embryos can be seen inside the eggs.

Opposite bottom A cuttlefish egg almost ready to hatch.

Below Embryonic development of an inshore squid (family Loliginidae), showing absorption of the external yolk sac as the embryo develops.

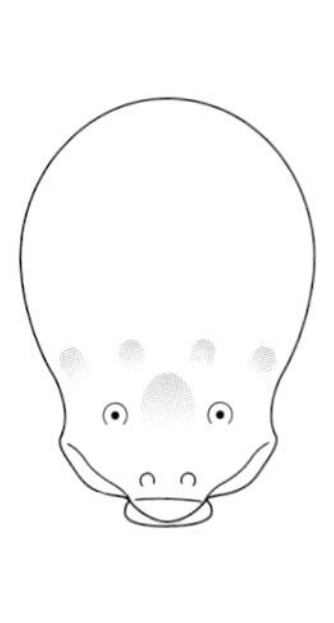

Early organogenesis

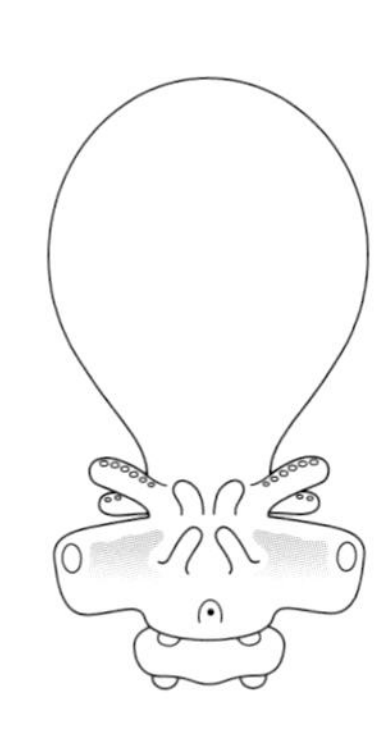

Early morphogenesis

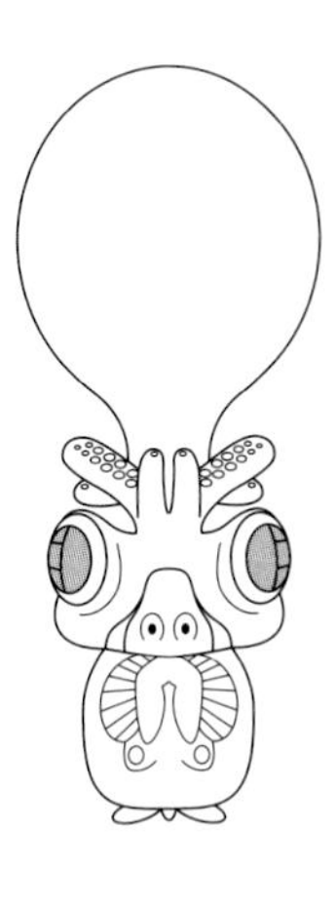

Middle morphogenesis

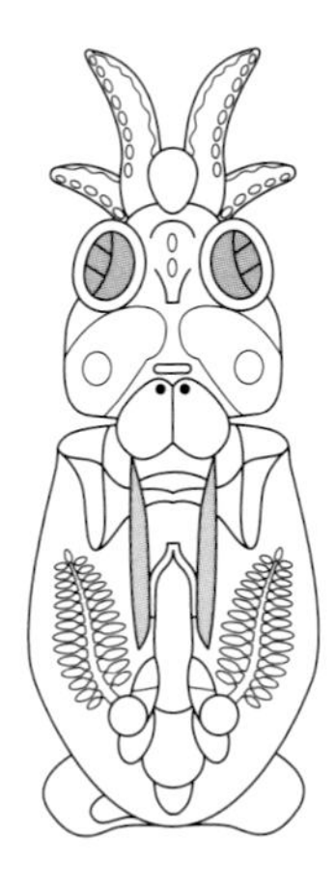

Almost fully developed

HATCHING & EARLY LIFE

Hatching occurs when the embryo is fully developed, and some cephalopods have a natural tranquilizer in the chorionic fluid to keep the advanced embryo quiet to prevent premature hatching, which could be harmful to survival. The fully developed embryos have organs that produce an enzyme that dissolves the chorion to help expel the animal out into the world.

There are two major types of hatchling: large hatchlings that are fully formed, sort of "miniature adults," or smaller hatchlings termed "paralarvae" because they still have recognizable features of the adults but have some different features that soon disappear with development. These latter are not considered true larvae, hence the modifier "para." There are ecological differences in the two as well: large adult-like hatchlings take up a benthic life like their parents, whereas small paralarvae become planktonic and drift for weeks before assuming adult body form and habit.

Below Argonaut eggs are very tiny. The eyes of the developing embryos can be seen inside these eggs.

Below right Octopod hatchlings. These will continue developing as planktonic paralarvae before settling onto the sea bottom.

Whether hatchlings are planktonic paralarvae or benthic juveniles is not consistent throughout all decapod families. The oegopsids all have paralarvae. The most extreme examples of this are in the squid family Ommastrephidae, for which fusion of the tentacles in the early post-hatching stage is a family characteristic, a so-called "rhynchoteuthion paralarva." However, benthic cuttlefishes and benthic bobtails generally hatch as benthic juveniles whereas the pelagic bobtails have planktonic paralarvae. In inshore squids, most genera have small eggs (less than 3/16 in (4 mm) long) and planktonic paralarvae but reef squids (*Sepioteuthis*) lay larger eggs that hatch into juveniles similar to the adults.

Among the benthic incirrate octopuses the size of the egg relative to the adult size determines whether benthic juveniles or planktonic paralarvae are produced. Thus, pygmy species of octopods may produce eggs that are quite small in absolute size but large relative to the small adults and hatch as crawl-away juveniles or "miniature adults," whereas a larger species that produces eggs the same size as those of the pygmy would go through a swimming paralarval stage after hatching. Closely related octopod species pairs that are otherwise almost identical are known to differ in egg size and developmental

Top right An advanced paralarva of a common octopus, *Octopus vulgaris*.

Middle right A "crawl-away" hatchling of the lesser two-spotted octopus, *Octopus bimaculoides*, leaving the egg mass.

Bottom right The paralarvae of the long-finned inshore squid, *Doryteuthis pealeii*, may disperse long distances from their hatching location on the sea bottom.

pattern. All pelagic incirrates have planktonic paralarvae, as do vampires. Cirrate octopods all produce large eggs; however, the proportional size of the large fins and short arms of hatchlings seem to indicate that they should be considered to be very large paralarvae. The large eggs of nautilids hatch juveniles that are very similar to the adults.

It is obvious that for benthic species with a planktonic stage a change in distribution occurs with ontogenic development. Ontogenic shifts in distribution also occur in pelagic species. The paralarvae often inhabit a different depth layer compared with older stages. The most typical pattern is that the paralarvae inhabit near-surface waters and then undergo an ontogenic descent, along with initiation of diel vertical migration. Other patterns of ontogenic migration are known, however, including ascent from deep waters and spreading from a narrow mid-depth zone.

In addition to shifts in distribution, ontogenic changes in feeding are also common, even for species with benthic juveniles. This generally involves a switch in prey type, not just prey size. Unfortunately, feeding by paralarvae is extremely difficult to study, so only the most general of patterns can be summarized. Internal yolk sustains the hatchling only for a few days. Feeding must begin before the yolk is completely exhausted but apparently does not usually begin immediately after hatching. Peak mortality in laboratory studies coincides with the time of yolk absorption, implying starvation as the cause.

AGE & GROWTH

HOW EASY IS IT TO DETERMINE THE life span of an animal? If you can observe the animal throughout its normal life, the task is easy. If you can hatch it and raise it until senescence, you know the animal's age and growth rates under the rearing conditions. If neither method of direct determination of age and growth is feasible, however, then indirect methods are required to infer these critical parameters. This, unfortunately, is the case for almost all cephalopods.

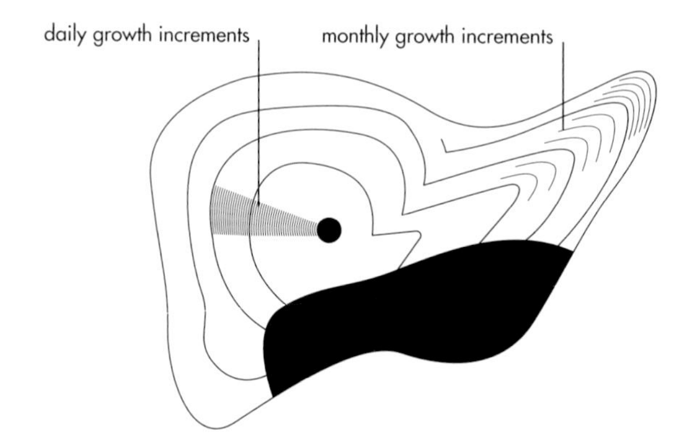

Top A statolith of squid *Doryteuthis opalescens* showing daily and monthly growth lines.

Above A standard way to determine cephalopod size is by measuring the dorsal mantle length, along the midline from the end of the mantle near the fins to the mantle edge by the head, visible here above the eye.

BEST GUESSES

Both tagging/recapture and growth in aquaria have been used effectively for species of *Nautilus*, and both methods generally agree on a life span of about one or two decades. Cohort analysis, an indirect method that has been successfully used in finfish fisheries science, was widely used until fairly recently for determination of age and growth in squids. This method has, however, been shown to be generally inappropriate for cephalopod studies. Alternative methods based on individuals rather than statistical trends indicate that growth rates in many cephalopod species are highly variable and therefore not appropriate for inferring growth from modal size shifts.

Squids conveniently have two structures, the gladius and the paired statoliths, in which growth is recorded as episodic marks while the structure increases in size. The numerous marks on both structures are assumed to record daily periods of growth, and this assumption has been generally substantiated for statoliths in a few coastal species. Less research has been devoted to the periodicity of growth lines on gladii. Formation of chambers in the cuttlebone of a cuttlefish may be similarly useful, although limited observations indicate that periodicity of chamber formation is less consistently daily than the growth marks on a gladius.

As mantle length is the standard measure of size in a squid and is directly related to the length of the gladius, growth marks on the gladius may seem to be the perfect index of squid age and growth. Unfortunately, total absolute age cannot be determined with this method. However, because growth lines on the gladius indicate the size of a squid, and differences in size separated by a known number of days can therefore be determined, growth rates can be calculated for individuals, as can changes in growth rate at different sizes. The assumption of daily periodicity in growth marks on gladii has rarely been tested.

Above Unlike their neocoleoid relatives that "live fast and die young," nautilids have several reproductive cycles during a life spanning many years.

SQUID STATOLITHS AS AGING TOOLS

Internal rings deposited as a statolith grows can be counted back to hatching. Thus the age can be determined for a squid of known size. If this is done for many squids of different sizes, growth curves and variability among individuals can be calculated. Daily deposition must be verified by chemically marking the exterior of the statolith in a living animal. The animal is then euthanized a known number of days later for dissection of the statoliths and counting the number of rings that have been deposited during the intervening time. This has been accomplished in aquaria for several coastal species, primarily inshore squids, lending support to the assumption of daily periodicity of deposition. It is possible, though, that the periodicity reflects artificial elements of the rearing methods. Validation of daily growth rings has been done in situ only once, with *Loligo reynaudii* in South Africa. Just counting the growth rings is difficult. Other than in paralarvae, the statolith must be carefully aligned and ground like a gemstone for the rings to be visible.

The evidence to date on these coastal species largely supports the daily deposition of statolith growth rings (with some variability toward the end of the life span). However, caution should be exercised in extending this assumption to cephalopods of other habitats, such as deepwater and polar regions, because periodic daily effects may differ among habitats or species-specific behaviors.

Octopods do not have gladii and their statoliths do not deposit clear growth rings. Attempts have been made, with limited success, to age species that have stylets by sectioning these remnants of the ancestral shell to count growth rings. Some naturally occurring chemicals apparently accumulate continuously throughout an animal's life and its age can be inferred from the concentration of these chemicals. This method may prove useful in the future for assessing the age of octopods and other cephalopods that lack clear growth rings on statoliths.

LIFE SPANS OF CEPHALOPODS

Whereas nautilids, which can be maintained in aquaria, are known to live for 15 years or more, available evidence for neocoleoids indicates life spans generally ranging from a few months for tiny species to fewer than 5 years at the most. This seems to be true even for very large species. The general pattern for most species is a life span of one or two years, with comparatively fast growth, especially in the large species. However, life spans of deep-sea species are essentially unknown; in a few known species embryonic development alone can take four years! As a result we are ignorant of the time needed to reach sexual maturation and spawning for many of the open-ocean and deep-dwelling cephalopods.

FAST GROWTH

Growth in weight by cephalopods in laboratory mariculture experiments provides a basic understanding of how fast some species can progress through the life cycle. Currently, the most widely accepted mathematical growth model is based on studies of incirrate octopods and inshore squids. Early growth is described by a logarithmic function and then, at some stage in

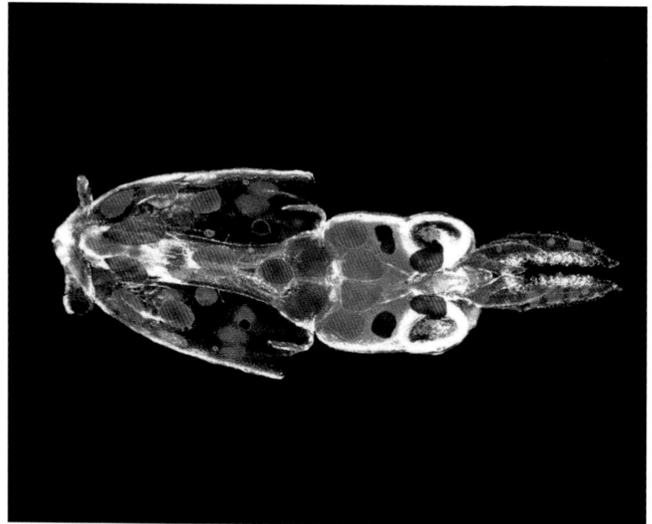

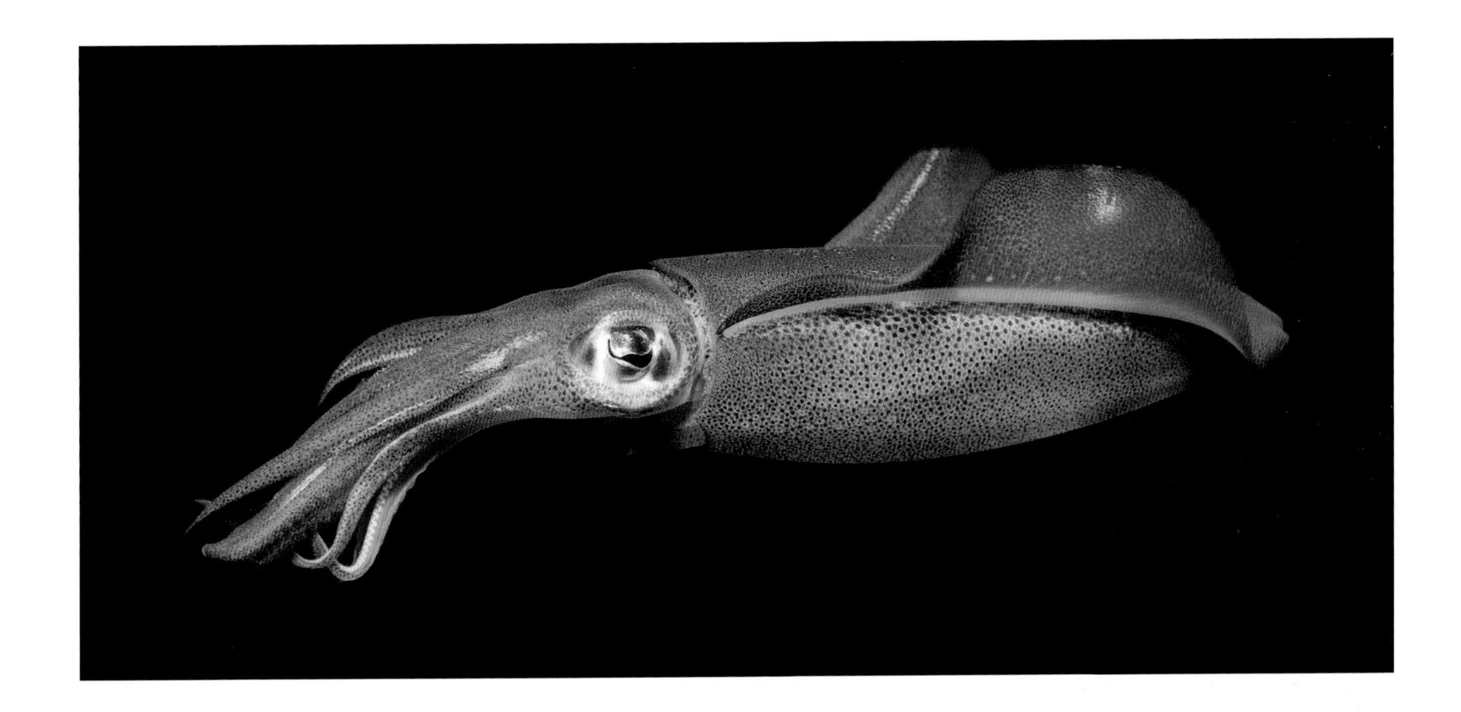

Opposite left Females of deep-sea warty octopods, like this *Graneledone*, care for their eggs for more than four years.

Opposite right Paralarvae like this common European squid, *Loligo vulgaris*, can grow to full size and maturity in about a year or less, depending on water temperature.

Above Most living cephalopods have life spans of a few months to a few years. For example, this reef squid can reach full adult size and maturity in less than six months.

the middle of the life cycle, later growth is better described as an exponential function. Both forms of growth are fast by any comparison—in early phases post hatching, inshore squids such as *Loligo* and *Doryteuthis* can double their weight in a week.

Measuring cephalopod growth by length is difficult and can be misleading. The most common metric is mantle length, and that can be fairly accurate for squids and cuttlefishes that have rigid structures in the mantle (gladius and cuttlebone, respectively). Yet using mantle length to measure octopods and bobtails is a poor metric because they lack a rigid structure to maintain consistent mantle length. Moreover, an octopod's size is determined more by its arms than by its mantle. Differences among species in relative proportions of arms, mantle, fins, and so on can be extreme. For cephalopods, it is generally easier to measure length than weight, so biologists have strived to provide length/weight measurements in key species (mainly fishery targets and biological model species) so that weights can be determined indirectly from length measurements.

Statolith ring counts in squids (generally one ring per day) combined with size measurements at the end of the life cycle are also good indicators of growth, and by this method the fast growth of many nearshore squid species has been determined. For example, large specimens of *Doryteuthis pealeii* with a mantle length of 10–12 in (25–30 cm) found off the northeastern USA are only nine to twelve months of age.

LOCOMOTION

MOST PEOPLE ARE AWARE THAT CEPHALOPODS swim by jet propulsion. Mantle jetting is not the only mode of locomotion for cephalopods, though. Locomotion in cephalopods is very complex. Most animals constantly use a combination of modes, whether an inshore squid flapping fins while jetting, a species of family Mastigoteuthidae undulating its fins to maintain a vertical posture while floating neutrally buoyant above the bottom, or an octopod rapidly switching from medusoid swimming to mantle jetting in the water column to crawling along the bottom.

Below In addition to crawling, benthic octopods can swim rapidly by jetting.

JET PROPULSION

Jetting is assumed to have evolved from an ancestor that swam similarly to a modern nautilid, with the head pumping in and out of the external shell like a piston in a cylinder. Evolution of a muscular mantle with compressible mantle cavity and flexible funnel has made the system much more effective. Of course, use of water jets for swimming is not unique to cephalopods. It is a common mode of locomotion for diverse groups of gelatinous zooplankton, including jellyfishes, and is used by other animals, such as scallops. What is unusual about jetting by most cephalopods is the ability to create high-pressure jets through a very directional tube, the funnel. This causes both opportunities and problems. The high-pressure jet allows the animal to accelerate rapidly. Because the funnel can be bent in any direction below the body, from straight ahead to straight

Above Some muscular oceanic squids, like these ommastrephids, can jet up out of the water and glide like flying fishes by spreading their fins and the membranes on their arms.

Above right Members of the family Mastigoteuthidae, like this *Magnoteuthis magna*, are typical squids in the deep oceanic waters of the bathypelagic.

astern, the cephalopod can easily jet forward or backward. There is a disadvantage to this system, though. The jellyfishes, for example, swim by slowly expelling a relatively large plug of water out of their bodies whereas cephalopods have to squeeze water through the constriction of a small tube—the funnel. Thus, mantle/funnel jetting by cephalopods is not very efficient in terms of the energy expended. Jetting is also a disadvantage energetically when compared with fishes, which swim with fins that displace water differently.

"Flying" is also accomplished by ocean squids such as *Todarodes pacificus*, which jet so forcefully that they are propelled out of the water, where they briefly continue to expel water from the mantle and concurrently extend their fins and arm membranes to act as wings to extend the glide path by many feet. This is used as an evasive escape from fast fish predators near the ocean surface.

SWIMMING LIKE A JELLYFISH

Both cirrate and incirrate octopods that have deep webs can use an alternative form of water jetting. By spreading their arms and webs and then drawing the arms together, they can eject a large volume of water at relatively slow speed, similar to the pulse of a jellyfish. This "medusoid" locomotion is an energetically efficient means of slow swimming and is widely used by octopods. Whether it is also used by the few squids with deep webs, such as *Histioteuthis bonnellii*, is unknown. The deep-sea squid *Mastigoteuthis magna* has been observed to roll its large fins together into a tube and slowly squeeze the water out of it to move, providing another kind of low-speed jet.

WHAT ARE THE FINS FOR?

How important the contribution of the fins to locomotion is for muscular squids is somewhat controversial. Most calculations of cephalopod energetics have been based on such squids and have assumed that the fins are of minor use in propulsion, functioning mostly for steering and stabilization, or to maintain thrust while the mantle sucks in new water. However, recent research on the estuarine squid *Lolliguncula brevis* has challenged this view, indicating that at some speeds the fins are important for both thrust and lift.

When observed *in situ*, squids, bobtails, cuttlefishes, and cirrates all appear to use their fins extensively. Indeed, cirrates and some deep-sea squids, like those in the families Mastigoteuthidae and Magnapinnidae, appear to rely almost exclusively on their fins for swimming. Species with long fins often use an undulating motion whereas those with shorter, broader fins flap them, although the distinction between flapping and undulation is not always clear. The rhythmic undulations traveling along the long fins on both sides of the body may be synchronized to swim forward or backward. Alternatively, these waves may travel along the fins in opposite directions on opposite sides of the body to rotate the body for tight turning, like a boat with twin propellers pushing in opposite directions to turn within its own length. Species with short fins also use them to swim either forward or backward.

BUOYANCY

The amount of swimming a cephalopod has to do depends on its lifestyle and buoyancy which, of course, are interrelated. Because muscle is denser than seawater, all cephalopods would be negatively buoyant, and would have to swim constantly or sink to the bottom unless they had a source of positive buoyancy to balance the negative.

Negative buoyancy is in fact the case for many muscular squids, such as families Loliginidae, Ommastrephidae, and Onychoteuthidae, and benthic octopods. However, species in many families have forms of flotation to counteract their sinking. The chambered phragmocones found in nautilids, sepiids, and the ram's horn squid are described in Chapter 1. Cuttlefishes pump gas in or out of their porous "cuttlebone" (which is calcium carbonate, not bone) to maintain neutral buoyancy with changes in depth. Nautiluses also maintain neutral buoyancy by pumping gas into or out of the many chambers in their coiled shell.

Many other species concentrate ammonia for buoyancy because it is lighter than the volume of seawater it displaces. Some pelagic octopods have even evolved a structure analogous to the swim bladder of a fish, using trapped air in a non-rigid pocket in the dorsal mantle cavity. Evidence exists that argonauts similarly trap air for flotation in the shell-like egg case.

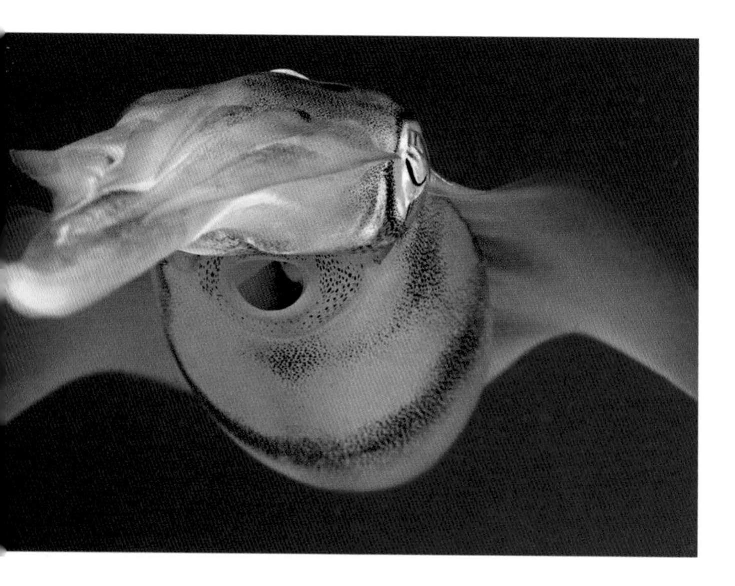

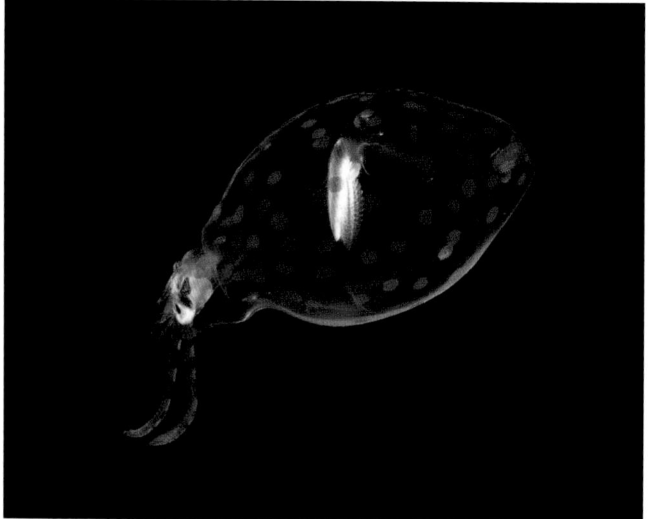

CRAWLING, WALKING, OR AMBLING INSTEAD OF SWIMMING

Of course a common means of getting about by benthic octopods is to crawl across the bottom. Usually they use all of their arms, reaching forward with alternating arms to grasp objects on the bottom with their suckers and pulling themselves forward in a highly coordinated manner. Whereas octopods are renowned for their well-developed brains, most of their nervous system is actually distributed in their arms. An octopod arm that has been severed or autotomized can crawl independently of the rest of the animal, useful for distracting a predator while the rest of the octopod departs. This seems to be a very advantageous strategy because a group of long-armed species in the family Octopodidae has evolved a transverse area of the arms near their bases where the muscle is modified to easily break away when stressed. These "autotomy planes" in the arm bases make each arm a bit like a lizard's tail that can break off and remain, wiggling, while its former owner makes its getaway.

Two octopod species (*Amphioctopus marginatus* and *Abdopus aculeatus*) have been described as having "bipedal" locomotion ("walking") because they crawl elevated on two arms while using the rest of their arms and marvelous skin to make themselves look very unlike an octopus. The colorful flamboyant cuttlefish *Metasepia pfefferi* also crawls slowly along the sandy bottom where it lives, using its ventral arms and two projections of the ventral mantle as if it were a four-legged bottom walker; this behavior has been referred to as "ambling."

Opposite left This reef squid, *Sepioteuthis*, is swimming with its fins as well as jetting through its funnel.

Opposite right Glass squids, like this *Cranchia scabra*, reduce their need to swim actively by maintaining neutral buoyancy using an internal chamber filled with ammonium chloride solution, which is less dense than seawater.

Below left Bipedal walking by the veined octopus, *Amphioctopus marginatus*, which here is using only its ventral pair of legs to move across the bottom.

Below right The flamboyant cuttlefish, *Metasepia pfefferi*, uses two bumps on the bottom of its mantle together with the two ventral arms to "amble" across the bottom as if it were four-legged.

CEPHALOPODS OF VARIOUS BIOMES

WE GROUP CEPHALOPODS ECOLOGICALLY BASED on where they live in the world's oceans. All cephalopods live in salt water, either marine or high-salinity estuarine environments. Except for the Black Sea, cephalopods are found in all oceans and seas, from pole to pole. There are cephalopods from the surface to the deep sea, but none has yet been found in the trenches that are the deepest marine biomes.

ESTUARIES

Although cephalopods have been around for many millions of years, they seem not to have been able to evolve the physiological mechanisms necessary to survive at salinities below about what you would get if you mixed equal amounts of ocean water and fresh water. For an oceanic animal to survive in half-strength salinity would be somewhat like a human surviving by drinking such salty water. This is the salinity found in the middle to lower reaches of most estuaries, depending on tidal cycle and the amount of rain that has recently fallen upstream. Only a very few inshore squid and bobtail species can tolerate this much osmotic stress. Almost all living cephalopods, and presumably their ancestors, are intolerant of salinity changes. The few cephalopod species that can live in estuaries allow the chemical composition of their body fluids to change, matching the osmotic concentration of the surrounding environment. Of such cephalopods, the best-studied is the thumbstall squid, *Lolliguncula brevis*, in the western Atlantic. This species occasionally reaches high abundances in lower estuaries and inner shelf areas. However, its maximum abundance in estuaries occurs after summers with little rainfall, when the estuarine salinity is high.

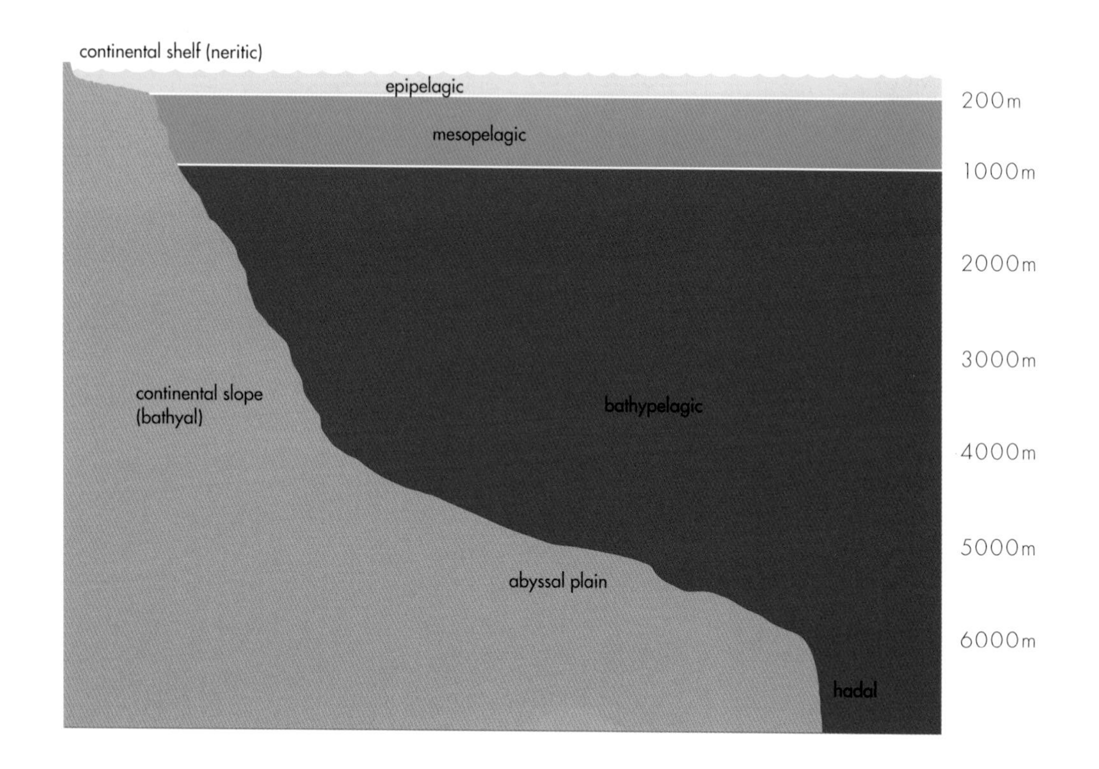

Left Diagram of major marine biomes. Cephalopods are known from all but the deepest (hadal) zone.

Left A tidal creek on the edge of the Chesapeake Bay, USA, in which the thumbstall squid *Lolliguncula brevis* can be found in waters with reduced salinity.

Above A pair of bigfin reef squid, *Sepioteuthis lessoniana*, swimming above a coastal bed of seagrass.

CONTINENTAL SHELVES

Nearshore habitats on the shelves include coral reefs, seagrass beds, kelp forests, tide pools, sediment plains, and rock reefs that collectively represent some of the most diverse biomes on Earth. Many species of cephalopods live in these habitats. The three most diverse cephalopod families, Sepiidae, Loliginidae, and Octopodidae, are common in these nearshore areas and across the shelf, although some species of each are found beyond the edge of the shelves in the waters of the continental slope for at least part of their life cycles. Many bobtail species also live on the shelves. Additionally, the oceanic squid family Ommastrephidae includes several species that conduct regular seasonal migrations onto continental shelves from nearby continental slope and oceanic areas. A major difference distinguishing these ommastrephids is that they produce pelagic egg masses whereas the other coastal families lay benthic eggs. All of the coastal cephalopods spend part of their time on the bottom, but get up and swim to some extent. Octopods, however, are more bottom-oriented, whereas loliginids spend much of their time swimming. Cuttlefishes and bobtails are intermediate between the benthic and pelagic extremes, spending part of their time cruising up in the water but much time either sitting on, or buried in, bottom sediments.

COASTAL DIVERSITY

The highest diversity of coastal cephalopods is found in tropical waters, particularly in the Indo-West Pacific region. This is the pattern seen in many types of organisms. In temperate and cooler waters, although there are fewer cephalopod species than in the tropics, coastal species sometimes attain very high abundances and cumulative biomass. Hence, some of these species are the targets of many of the largest cephalopod fisheries. One peculiarity of the distribution of tropical and temperate cephalopods is the absence in the Americas of cuttlefishes and one sepiolid subfamily, the Sepiolinae, discussed in Chapter 2.

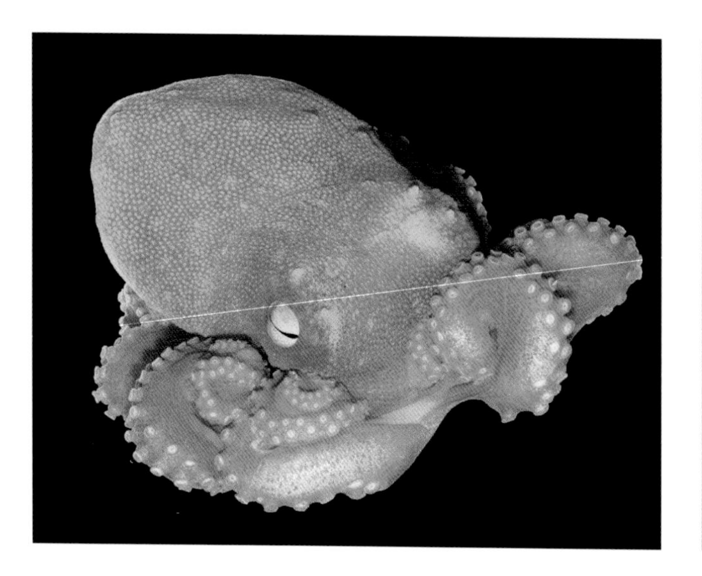

Above left As its name suggests, *Bathypolypus arcticus* is found in the Arctic Ocean.

Above This Antarctic knobbed octopod, *Adelieledone polymorpha*, is one of many species of incirrate octopods that live around Antarctica.

The diversity of coastal cephalopods in polar regions is very different between the Arctic and Antarctic. The Arctic has a few species of incirrate octopods and bobtail sepiolids. Until recently, the coastal cephalopods of the Antarctic were thought to comprise just a few morphologically variable species of incirrates. However, detailed morphological and genetic research has shown that the incirrate diversity in the Antarctic had been underestimated by an order of magnitude. Although it is possible that a similar underestimation has occurred in the Arctic, low diversity in the Arctic and high diversity in the Antarctic fit the pattern for other major evolutionary lineages such as fishes and isopod crustaceans.

OPEN OCEAN—THE VERTICAL & TIME DIMENSIONS

Let's skip momentarily over the continental slopes because they are easier to understand as an interface between the coastal and the truly oceanic biomes. The open ocean is a four-dimensional system, strongly structured by depth and time. The depth distribution of most pelagic animals changes with time of day and season. These movements are the true nature of pelagic animal life. Before we consider migratory patterns, though, let's begin with a static view of the vertical structure of the water column. The pelagic environment can be divided into layers. Although some organisms are confined to the air–water interface, there are probably no such cephalopods. The small bluish-colored squid *Onykia carriboea* has been referred to as a surface dweller, but is actually the young of a deep-living species. The paralarvae of some species, such as *Sthenoteuthis oualaniensis*, are caught in much higher numbers in surface nets than deeper in the water, but their later stages are definitely not confined to the surface. The layer of open ocean closest to the surface is the epipelagic, which encompasses approximately the upper 660 ft (200 m) of water. Although many cephalopods live in the epipelagic, only a few squid and incirrate octopod species spend their entire lives there.

Left A juvenile histioteuthid. It is morphologically intermediate between a paralarva, found consistently near the surface in oceanic waters, and an adult, which would migrate vertically every day from the twilight zone of the mesopelagic in daytime to the near-surface epipelagic at night.

Above The greater argonaut, *Argonauta argo*, lives near the surface in warm oceanic waters around the world.

The mesopelagic zone, from about 660 to 3,300 ft (200 to 1,000 m) depth, is the center of oceanic-squid diversity, at least during daytime. Several pelagic incirrates and a few bobtails are found here as well. Mesopelagic squids include the small but abundant and beautifully bioluminescent enoploteuthid group of families and family Histioteuthidae, known as strawberry or jewel squids. Most of the muscular oceanic squids, such as ommastrephid flying squids and gonatid clubhook squids, spend their days, but not their nights, in the mesopelagic. At night, they move closer to the surface.

The bathypelagic realm extends from about ½ to 2½ miles (1 to 4 km) depth and is beyond both the maximum penetration of sunlight at noon in clear oceanic waters and the maximum depth of the permanent temperature gradient. The residents of these vertical zones are not as easily categorized by depth zone as was once thought. For example, vampires are found in the upper bathypelagic and lower mesopelagic.

The bathypelagic realm is by far the largest percentage of living space on earth, and most in need of exploration. Animals of the bathypelagic tend to be gelatinous and are not strong swimmers. Among the typical cephalopods of the bathypelagic are squids such as family Mastigoteuthidae and other members of the chiroteuthid group of families, and cirrate octopods of family Cirroteuthidae.

When you get close to the deep-sea floor, within about a thousand feet, you encounter the "benthopelagic" fauna, which is pelagic but strongly associated with the bottom. Animals here live in an enriched (relative to the overlying water) near-bottom zone known as the benthic boundary layer. Some of the bathypelagic cephalopod families include species that are benthopelagic in the abyss, including species of squid family Mastigoteuthidae and two families of cirrate octopods. These benthopelagic animals are distinguished from the benthic and demersal fauna because they seldom, if ever, sit on the bottom. True bottom dwellers of the abyssal deep sea are strongly dominated by octopods, both cirrate and incirrate. Whereas cirroteuthid and stauroteuthid cirrates spend most of their lives up in the water, apparently coming to the bottom only to attach their eggs, the more speciose family Opisthoteuthidae spend much of their time sitting on the bottom. The taxonomy and evolutionary relationships of abyssal and bathyal incirrates have been very confused. Evolution of many deep-sea animals has involved losses of anatomical structures. Many species of deep-sea benthic incirrate octopods with bi-serial suckers have historically been lumped together based on the lack of an ink sac. Loss of the ink sac in the darkness of the deep is easily convergent and classification based on this probably does not reflect evolutionary history.

The really deep waters of the oceanic trenches, from about 3¾ miles (6 km) deep to the deepest part of the ocean at about 6¾ miles (11 km), occupy only a very small percentage of the ocean. No cephalopod species are currently known from these "hadal" environments. However, trenches are the most difficult environments of the planet to explore and this lack of cephalopods may be an example of how little is actually known about life in the deep sea.

As with the coastal cephalopods, the diversity of polar oceanic cephalopods appears to be much lower in the Arctic than in the Antarctic. Whereas only one oceanic squid species and one cirrate octopod species have been commonly found in the oceanic waters of the Arctic, dozens are known from the Southern Ocean, including the endemic family Psychroteuthidae.

WHERE CONTINENT MEETS OCEAN

Now back to the continental slopes and rises, or "bathyal" biomes, with bottom depths ranging from the edge of the shelf at about 660 ft (200 m) to about 3,300 ft (1,000 m) (continental slope) and on out to about 11,500 ft (3,500 m) (continental rise). With a few exceptions, the slope and rise environments are essentially an interface between the coastal and the truly oceanic. When sampling the bathyal, you can encounter families typical of the shelf (such as Loliginidae and Sepiidae), the epipelagic (such as Ommastrephidae and Onychoteuthidae), mesopelagic (Histioteuthidae and the enoploteuthid families), bathypelagic (Mastigoteuthidae and vampires), and the abyss (*Grimpoteuthis* and *Muusoctopus* spp.).

Additionally there are evolutionary groups, such as the incirrate octopod genus *Bathypolypus*, that are best characterized as typically bathyal. Nautilids, which live on deep reef faces to depths of about 2,000 ft (600 m), should be considered part of this fauna. As you might expect, bathyal cephalopod assemblages can be diverse, complex, and dynamic.

Opposite Members of the family Mastigoteuthidae, like this *Mastigoteuthis agassizii*, are typical squids in the deep oceanic waters of the bathypelagic.

Above left Some finned "dumbo" octopods, like this cirroteuthid, leave their benthic eggs to spend their lives swimming in the deep bathypelagic.

Above This deep-sea squid, *Chiroteuthis calyx*, holds its tentacles in a special sheath formed by the keels on the very large ventral arms.

The complexity of the bathyal assemblages became more apparent with the discovery of what is called the mesopelagic boundary community. This is a group of species, including cephalopods as well as fishes and shrimps, in families that have long been considered to be oceanic/mesopelagic. The boundary species, however, are strongly associated with slopes. For many boundary species, closely related sister species are known that are truly oceanic, so within a genus there may be one or several pairs of very similar species, one of which is boundary and the other oceanic.

MIGRATION

IN CONTRAST WITH A STATIC SNAPSHOT OF cephalopod distribution, the real situation is greatly complicated by movements: daily, seasonal, and shifts among life-history phases. Diel vertical migration must be a strong advantage because many animals undertake it. Seasonal migrations may be for feeding or reproduction. Whereas the daily and life-history shifts in distribution are mostly vertical movements of thousands of feet, seasonal changes can cover large horizontal distances.

DAILY MOVEMENTS

Daily, or diel (not, as is often used, diurnal, which means during the daytime only, in contrast to nocturnal) migration is primarily vertical in direction. Diel vertical migration is most pronounced in the mesopelagic, where the cephalopods participate in the largest migration on earth. Most diel migration is upward with decreasing light levels in the late afternoon and evening and downward as light increases after sunrise. The distance of these movements varies among species but can be extensive, up to 3,300 ft (1,000 m) or more from the lower mesopelagic and upper bathypelagic all the way to the surface for some species.

In addition to the common upward movement at night, evidence indicates that some species occupy a preferred depth during the day but spread both upward and downward at night. Furthermore, the mesopelagic boundary community is again unusual because, in addition to moving up and down on a diel basis, they shift horizontally up and down the sloping bottom and are therefore closer to shore at night than during the day.

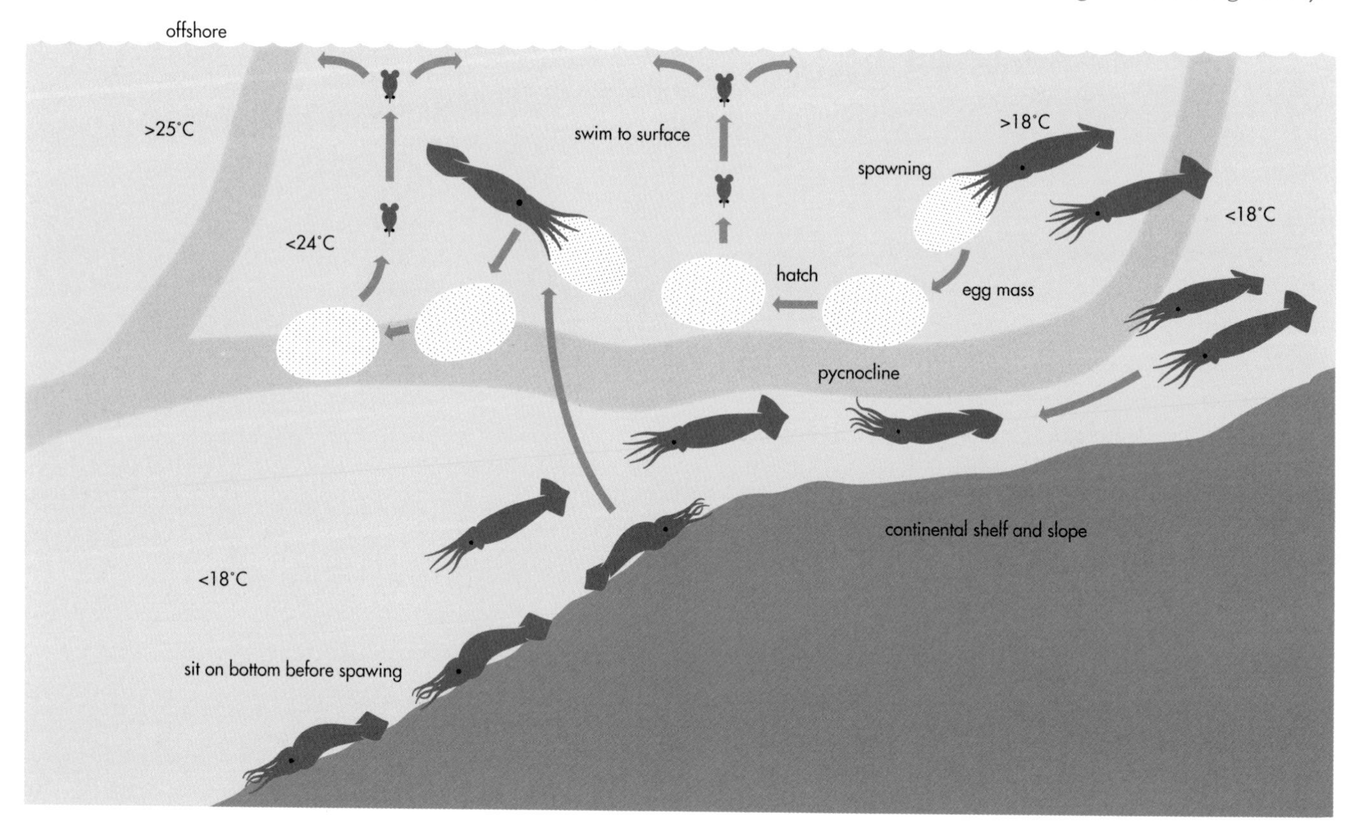

Nautilids undergo a similar daily migration up and down the deep reef face. Diel vertical migration is not just an open-ocean phenomenon. Some coastal species tend to sit on, or bury in, the bottom during the day, then spread upward throughout the overlying water column at night. This is why inshore squids are easily caught by bottom trawls during the day, but jigged or gathered into surface nets at night.

MOVING INTO NEW NEIGHBORHOODS

In addition to vertical migrations, which tend to be diel, some cephalopods also undertake horizontal migrations, which tend to be seasonal in nature. Seasonal migrations of cephalopods are not as well documented as the diel migrations and are best known for the species that are fished commercially throughout most of the year. They seem to move for two principal reasons: feeding and spawning. These moves can result in dense seasonal aggregations that can then be targeted by fisheries.

For coastal species, horizontal migrations may be inshore/offshore, or along the coast (north–south along the east coast of North America, for example), or a combination of both. Species such as the inshore squid *Doryteuthis pealeii* off northeastern North America and the cuttlefish *Sepia apama* off southern Australia migrate inshore to aggregate in preferred spawning areas. Conversely, the short-finned squid *Illex illecebrosus* seems to migrate inshore to feed and then offshore to spawn in more oceanic conditions. Additionally, increasing evidence, including tag-and-recapture studies, indicates that species of family Ommastrephidae like *Illex illecebrosus* and

Opposite In addition to daily vertical migration, the distributions of many species change throughout their life cycles, as illustrated here for the Japanese flying squid, *Todarodes pacificus*.

Top right The giant Australian cuttlefish, *Sepia apama*, migrates inshore to aggregate in preferred spawning locations.

Bottom right Some squids, like *Todarodes pacificus*, migrate very long distances toward the equator to spawn in the proper oceanic conditions.

Right In many deep-living oceanic species, like this *Leachia*, the early stages are found closer to the surface than adults.

Opposite top Some oceanic species, such as the diamondback squid (*Thysanoteuthis rhombus*), migrate great distances during their life cycles.

Opposite bottom Young stages of *Vampyroteuthis infernalis* are found deeper in oceanic waters than the adults.

Todarodes pacificus, living in regions where there are strong western boundary currents (Gulf Stream and Kuroshio Current), migrate toward the equator to spawn their pelagic egg masses in the edge of the current. The eggs and subsequent paralarvae drift in these strong currents toward colder water. When they become juveniles, they develop the ability to swim inshore to feeding areas, completing a very large-scale life-history loop in about a year.

Large-scale geographic seasonal migrations by truly oceanic species, usually north–south in direction, have only been found so far in large muscular epi/mesopelagic squids such as families Ommastrephidae and Thysanoteuthidae that are commercially harvested. It is possible that other oceanic species have seasonal migrations. For example, large aggregations of the epipelagic incirrate octopod *Argonauta* ("paper nautilus") have often been reported and may result from spawning migrations.

GROWING UP & MOVING AWAY

The geographic migration of *Illex illecebrosus* described above is ontogenic as well as seasonal in nature. In other words, different life-history stages are found in different places. Ontogenic migrations may also be vertical. Benthic species with planktonic paralarvae that then settle to the bottom as they develop are obvious ontogenic migrators. Species living their entire lives in the water column may undergo ontogenic migrations as well. Often this is found in squids such as family Cranchiidae, in which the advanced juveniles and adults live in the lower mesopelagic or bathypelagic but the paralarvae are found in the epipelagic, presumably because there is more food there. Such patterns may occur because the eggs are buoyant and float to the surface, or because the eggs are actually spawned at the surface, or because the hatchlings are attracted to light or swim toward the surface against gravity.

In a few cases, such as the vampire *Vampyroteuthis infernalis* and the weird ram's horn squid *Spirula spirula*, the youngest stages appear to be located deeper in the water than the adults, perhaps because the eggs are either negatively buoyant or actually attached to the bottom in deep water, as with pelagic cirrate octopods that attach their eggs to deep-sea corals or sponges.

Ontogenic shifts in distribution may be gradual, slowly shifting throughout much of the paralarval development period, or quite abrupt, the paralarvae undergoing rapid changes in appearance accompanied by a concurrent rapid descent through the water column. Vertical ontogenic shifts are also complicated by the development of diel migration with growth.

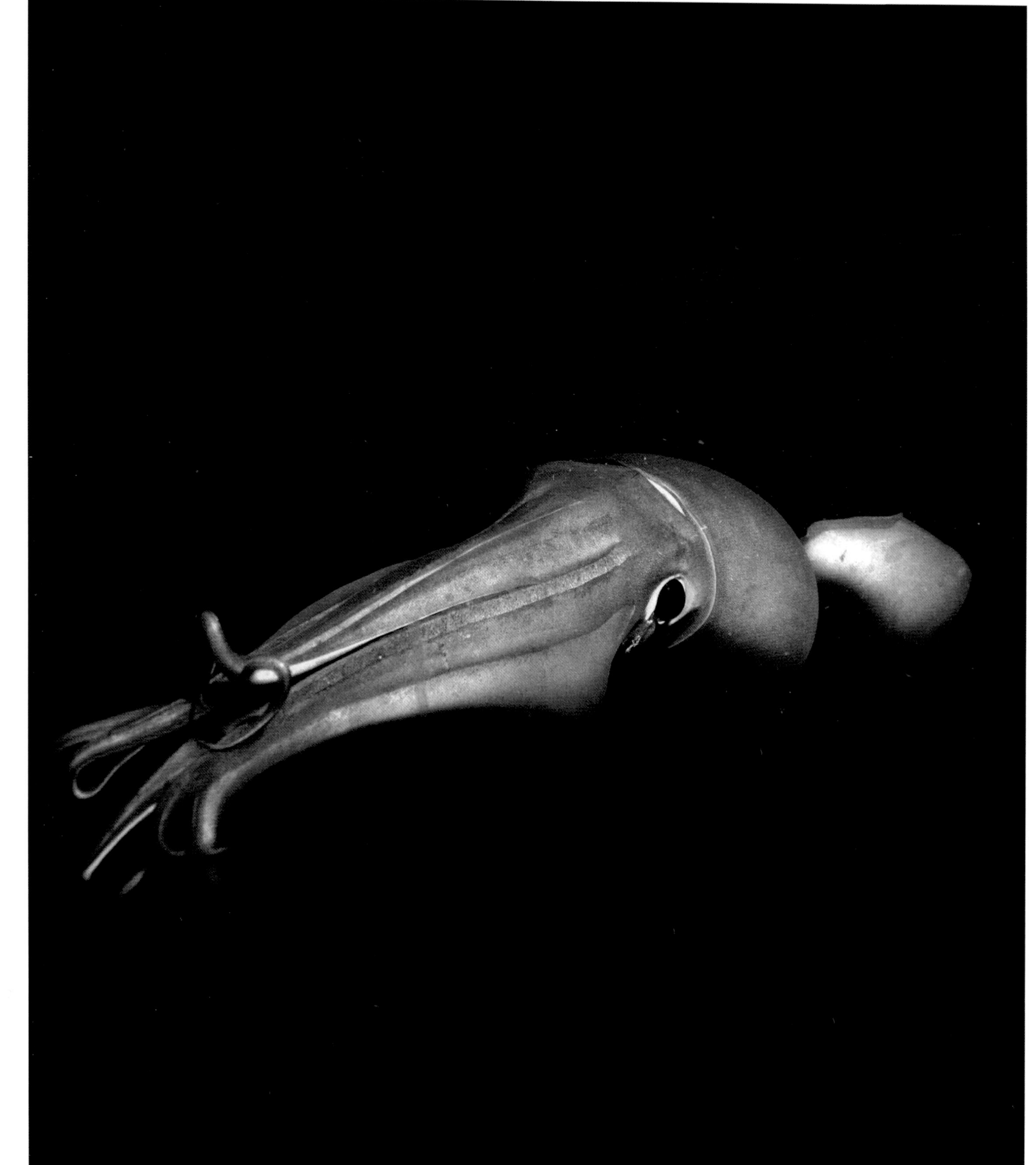

HUMBOLDT SQUID

Dosidicus gigas

FAMILY	Ommastrephidae
OTHER NAMES	Jumbo flying squid
TYPICAL HABITAT	Eastern Pacific Ocean, including areas with oxygen-minimum zones
SIZE	Mantle length 48 in (120 cm)
FEEDING HABITS	A voracious predator of other nekton species
KEY BEHAVIORS	Capable of tolerating low-oxygen conditions

THE HUMBOLDT SQUID IS SOMETIMES CONFUSED with the giant squid because it is the largest ommastrephid species and is often seen by people within its geographic range. Humboldt squids are fished both commercially and recreationally and mass strandings are not unusual. Therefore, unlike true giant squid, they are commonly encountered by people. Three separate populations have been suggested based on size at maturity. However, an alternative explanation is that these size differences are a growth response to varying environmental conditions such as the El Niño/Southern Oscillation phenomenon. From a core distribution in equatorial and tropical areas of the eastern Pacific Ocean the species sometimes spreads both to the north, as far as southern Alaska, and south, to Chile. Humboldt squids use metabolic suppression to survive within the layer of the water column with dissolved-oxygen concentrations too low for most other large swimming predators. This gives the squids both a competitive advantage over animals that eat the same prey as them and a refuge from large fishes that feed on squids.

FIERCE REPUTATION

Large, muscular, and aggressive, Humboldt squids form schools that have been documented to attack scuba divers. They have a reputation as “red devils” that many people fear, although the reputation has been challenged by experienced divers who have worked with them.

SPOTTED AT THE SURFACE

The Humboldt squid is a large species that migrates vertically from depths sometimes exceeding 3,300 ft (1,000 m) to near the surface at night. These visits to the surface, together with the large size, make it a popular subject for nature documentaries on television.

GREATER ARGONAUT

Argonauta argo

ARGONAUTS ARE PELAGIC OCTOPODS WITH STRONG sexual dimorphism. Female argonauts look much like familiar bottom-dwelling octopuses but they spend their entire lives swimming up in the water column. The males are much smaller and look like octopus paralarvae except for their peculiar arms. Even the smallest males have a developed hectocotylus stored in a pouch under the eye. During mating, the hectocotylus is released from the pouch and detaches from the male, carrying a spermatophore with it. Mated females carry the detached hectocotylus with them. Because when a mated female is captured in a net the hectocotylus can sometimes be seen wiggling, it was thought to be a parasitic worm and was named as a separate genus, *Hectocotylus*.

AN EGG CASE FOR A HOME

There are many peculiar features about argonauts. The female secretes and lives in a structure that appears to be a shell, but is not. Because of this structure, argonauts are sometimes called paper nautiluses. This calcium carbonate structure is secreted by flaps on the tips of the two dorsal arms. Because the female lays her eggs in it, the structure is really an egg case. Because she lives in it, early observers speculated that it was a boat and that the flaps were used as sails, hence the name argonaut. Recently, females have been shown to trap air at the surface in the egg case and swim down to a depth where the compressed air creates neutral buoyancy for the argonaut.

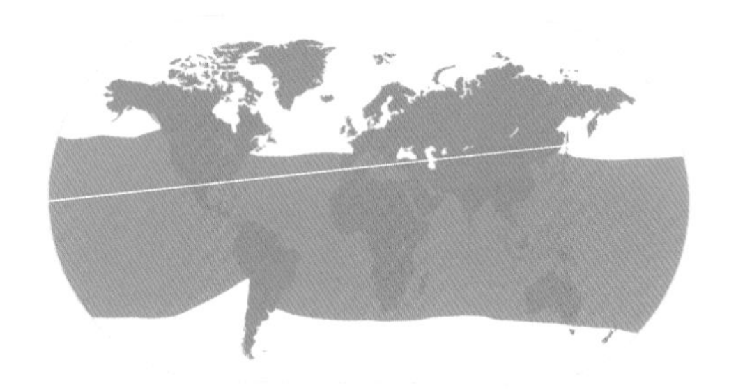

FAMILY	Argonautidae
OTHER NAMES	Paper nautilus
TYPICAL HABITAT	Warm near-surface waters of the open ocean
SIZE	Shell-like egg case to 12 in (30 cm) diameter
FEEDING HABITS	Have been found attached to, and perhaps eating, jellyfishes
KEY BEHAVIORS	Females brood their eggs while living in an egg case

PRECIOUS SHELL

The "paper nautilus shell" is prized by shell collectors and sometimes found washed up on beaches. Pieces of this fragile structure are often found in the stomachs of predators such as tunas.

STRIPED PYJAMA SQUID

Sepioloidea lineolata

FAMILY	Sepiadariidae
OTHER NAMES	Striped dumpling squid, striped bottletail
TYPICAL HABITAT	Very shallow sand and rubble off Australia
SIZE	Mantle length 2¾ in (7 cm)
FEEDING HABITS	Ambush passing prey while hiding buried
KEY BEHAVIORS	Bury in sand during daytime with only eyes exposed

LIKE OTHER SPECIES IN THIS FAMILY, the striped pyjama squid is short and broad with a fixed color pattern and lives on the bottom in shallow warm water. The common name derives from the prominent brown or black stripes on a white background over the head, mantle, arms, and bases of the fins. Such prominent coloring may advertise a poisonous quality to deter predators. Slime produced by the skin has been speculated to be toxic. The round, large (compared with adult size) eggs are laid in rock crevices. The young hatch out as fully formed miniatures, complete with stripes like those of the adults.

A SQUID WITH EYEBROWS

The fringes of finger-like papillae at the dorsal edge of the mantle opening over each eye form a structure reminiscent of eyebrows.

SMALL AND STRIPED

Unlike the rapidly changing color patterns of many cephalopods, the stripes on these cute bobtails are permanently fixed. Although small, bobtails have become popular for exhibit at public aquaria.

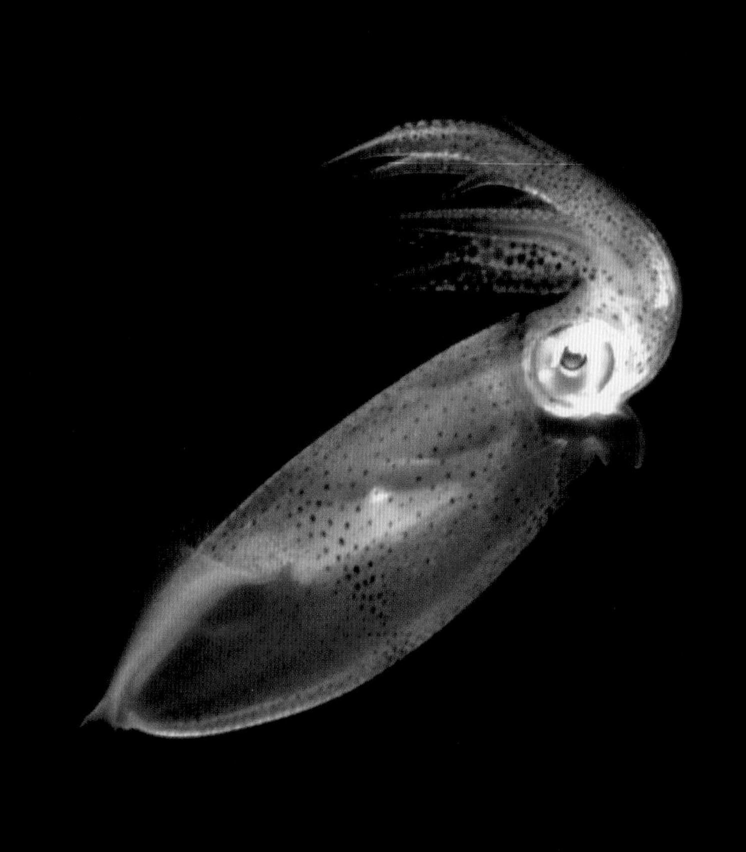

THUMBSTALL SQUID

Lolliguncula brevis

FAMILY	Loliginidae
OTHER NAMES	Brief squid, bay squid
TYPICAL HABITAT	Shallow coastal and lower estuarine waters
SIZE	Females to 4¾ in (12 cm) mantle length, males to 3⁵⁄₁₆ in (8.5 cm) mantle length
FEEDING HABITS	Fishes and shrimps
KEY BEHAVIORS	Attach egg masses to hard substrate, including artificial structures

THIS SMALL COASTAL MYOPSID SQUID IS sometimes very abundant and is certainly important in the food webs of local ecosystems. It has often been overlooked, though, in studies of estuarine and coastal nekton, which typically focus on either fishes or crustaceans. Much knowledge of marine biology often derives from fisheries. Because of their small size, thumbstall squids are not targets of commercial fisheries as are many other species of the family Loliginidae. They are caught as by-catch in shrimp and finfish fisheries and either consumed locally or used as bait for fishing. *Lolliguncula brevis* is reported over a very great latitudinal range along the western margin of the Atlantic Ocean. Evidence is accumulating that this is another example of a complex of very similar species.

AN ESTUARINE CEPHALOPOD

Almost all cephalopods require the full saltiness of seawater. However, *Lolliguncula* can live where ocean water is diluted by fresh water to as low as half-strength as well as in waters up to full ocean salinity. It can also tolerate a broad range of temperatures and dissolved-oxygen concentrations. This physiology allows the species to inhabit the lower reaches of estuaries throughout its geographic range, where other squids cannot go. Because it also spawns near the mouths of estuaries, paralarvae of the thumbstall squids are typically the only cephalopods found in estuarine zooplankton.

ESTUARY DWELLER

Like its loliginid relatives, thumbstall squids exhibit a suite of postural and chromatic behaviors even though the estuarine waters in which they live often have low visibility. These behaviors, however, have not been studied in as much detail as have those of some other loliginids. Thumbstall squids have proven to be a very convenient species for experiments on squid hydrodynamics.

AGASSIZ'S WHIPLASH SQUID

Mastigoteuthis agassizii

MASTIGOTEUTHIDS ARE THE MOST TYPICAL squids of the deep sea. They are called whiplash squids because of their unusual tentacles, which lack conventional tentacular clubs and look somewhat like a pair of whips. The tentacles are long, thin, and white, without much musculature, and are almost covered by many, very tiny suckers. The tentacular suckers of this species are too small to see without magnification. This arrangement of many minute suckers makes the tentacles sticky. Mastigoteuthids drift in the deep sea with their tentacles dangling downward. Presumably, if potential prey in the darkness bumps into a tentacle, it will stick and then the tentacle can be drawn slowly into the arm crown.

GUESSING ABOUT FUNCTION CAN BE MISLEADING

Cephalopod biologists who examined *Mastigoteuthis* that had been caught in nets realized that the ventral pair of arms was very large and contained ammonia, which caused them to float in seawater. Because of this, biologists speculated that in its natural habitat the squid floats with its head and arms upward. When they were finally seen alive in the deep sea, their posture was the opposite of the speculation. They use their fins to actively maintain position with the arms downward and held separated. Each tentacle is held in a sheath on the corresponding ventral arm from which it dangles awaiting contact with prey.

FAMILY	Mastigoteuthidae
OTHER NAMES	Red deep-sea squid
TYPICAL HABITAT	Truly deep sea (bathypelagic), often drifting very near the bottom
SIZE	Mantle length to about 5 in (13 cm)
FEEDING HABITS	Very small crustaceans
KEY BEHAVIORS	Drifts with sticky tentacles dangling down to ensnare prey

TENTACLE CONTROL

Mastigoteuthids can control their weak but very sticky tentacles by holding them in a groove, or "tentacular sheath," along the side of the enlarged ventral arms. When fishing for prey, the white tentacles dangle downward from the tentacular sheath, but when swimming the squid can withdraw the tentacles entirely within the sheaths. The small dark dots scattered on the skin are photophores.

PACIFIC WARTY OCTOPUS

Graneledone pacificus

FAMILY	Megaleledonidae
OTHER NAMES	Pacific deep-sea octopus
TYPICAL HABITAT	Found on hard substrate at depths of ½–1¼ miles (1–3 km)
SIZE	Mantle length to 6 in (15 cm)
FEEDING HABITS	Unknown
KEY BEHAVIORS	Very long-term brooding of large eggs

SOME CONTROVERSY REMAINS ABOUT WHETHER this nominal species comprises a single biological species widespread across the north Pacific or two geographically separated species, *Graneledone pacificus* and (farther north) *Graneledone boreopacificus*. All species of *Graneledone* appear "warty" because they are largely covered by cartilaginous tubercles. The function of the tubercles is not known but is likely to be protective in some way. Warty octopods are commonly encountered on rocky-bottom terrain along continental slopes and seamounts in the deep sea. Like many deep-sea octopods, they lack an ink sac; this lack has now been shown to be convergent in several evolutionary lineages of deep-sea octopods. The single series of suckers on each arm is an ancestral characteristic.

VERY LONG-TERM MATERNAL CARE

During a periodic deep-sea survey by the Monterey Bay Aquarium Research Institute using a remotely operated vehicle, a female warty octopod was found brooding a clutch of eggs attached to a rock. This fortunate circumstance allowed the scientists to revisit the female and her eggs periodically throughout the next few years. From the series of dives, they inferred that this species broods its eggs, which are close to the largest known for invertebrates, for at least 53 months. This was also the first direct measurement of longevity in a deep-sea cephalopod, much longer than the life span of several months to a couple of years generalized from shallow-water species.

ONE SERIES OF SUCKERS
Many cartilaginous tubercles are obvious in the skin of this "warty" octopod. The single series of suckers on the arms can be confusing when watching a live animal because when the arm is contracted the suckers move into a zig-zag configuration, which is easily confused with the two series seen in many other octopod species.

BALLOON DUMBO OCTOPUS

Stauroteuthis syrtensis

STAUROTEUTHIS SYRTENSIS IS THE BEST KNOWN OF the cirrate octopods that spend all of their time swimming. They have been seen from deep-sea submersibles many times but have never been observed resting on the bottom. They sometimes drift close to the bottom with their arms spread like an umbrella and it is believed they visit the bottom to deposit eggs on deep-sea corals or sponges. This species has an extensive and complex system of webs connecting its arms. When it is disturbed—by a submersible, for example—it spreads its arms and then draws the arm tips together. This behavior traps a large volume of water and inflates the webs, making the animal look like a pumpkin-shaped balloon. There are several possible explanations for this behavior in the permanent deep-sea darkness although no evidence is available to choose which explanation is most reasonable. The inflation may change its acoustic characteristics as a defense against deep-diving toothed whales; or, as the inflated octopod resembles a very large jellyfish, it might be pretending that it has jellyfish stingers.

GLOWING SUCKERS

Light-producing organs have been found in very few octopod species. *Stauroteuthis syrtensis* has what may be the strangest cephalopod photophores. Evolution has transformed its suckers into light organs. This was discovered accidentally when a bioluminescence researcher happened to catch a live specimen and shook it in a shipboard darkroom, just to see whether anything lit up.

FAMILY	Stauroteuthidae
OTHER NAMES	Glowing dumbo octopus
TYPICAL HABITAT	Deep open water near continental slopes and seamounts
SIZE	Mantle length to 14 in (35 cm), total length above $3^1/_3$ ft (1 m)
FEEDING HABITS	Feeds on small crustaceans, possibly using mucus to trap them
KEY BEHAVIORS	Swims with fins or with arms and web, but not by conventional cephalopod jet propulsion

DOUBLE WEBS

The very thin, membranous web of this finned octopod is not connected directly to the arms. Rather, it is connected by another set of membranes called the secondary web. This increases the volume of water that can be trapped in the inflated web system when the animal is disturbed.

BEHAVIOR, COGNITION & INTELLIGENCE

DECISION-MAKING: BIG BRAINS ENABLE COMPLEX BEHAVIORS

CEPHALOPODS ARE RENOWNED FOR BEING "the smartest invertebrate animals" but how do we assess this in such an alien body form and then substantiate it with scientific evidence? One way to judge cognitive adaptability is to test and measure a cephalopod's capability for rapid and diverse decision-making in a range of complex real-life scenarios. A large brain can facilitate these processes. Unlike most other animals, cephalopods augment much of their behavior and decision-making by changing their appearance—often very dramatically—for defense, feeding, and reproduction. Thus "rapid adaptive coloration" for camouflage and communication is considered a hallmark of soft-bodied cephalopods.

Decision-making is complex: it depends upon the integration of multiple sensory inputs, and cephalopods are well endowed with vision, taste, touch, smell, and hearing. Then a course of action must be determined, and here is where cephalopods excel compared with other invertebrate animals: their agility, speed, and color-changing abilities provide them with multiple options for action. Their huge brain, with 34 lobes and millions of tiny neurons, enables an enormous repertoire of behaviors for all occasions.

FAST LEARNERS

There has been a great deal of laboratory experimentation on visual discrimination learning in *Octopus vulgaris*. Most of this occurred in the 1950s and 1960s at the Stazione Zoologica in Naples, Italy. Octopuses can accomplish associative learning of some tasks in just a few trials. Octopuses can learn more than

Below *Octopus vulgaris* is extremely alert all the time as it forages amidst complex environments that have many predators.

Above left *Octopus cyanea* in a forward swimming posture as it moves across a Pacific coral reef.

Above *Octopus cyanea* crawling with stealth across a coral reef.

one visual discrimination at the same time. Octopus touch learning (via the suckers in their eight arms) is also rapid. Their ability to solve tasks such as serial reversal learning may relate to a flexible feeding strategy ("win–stay but lose–shift") that depends upon learning and might be useful to a predator searching for patchily distributed prey. There are equivocal results from testing observational learning from conspecifics in octopuses and cuttlefishes, although most authors estimate that octopuses may be able to accomplish some sort of improvement in performance by watching other animals, such as their prey, in the wild. These types of experiments remain to be undertaken. In addition, both short-term (one hour) and long-term (several months) memory are well documented in octopuses and cuttlefishes. Recent studies have provided good evidence for episodic-like memory in cuttlefishes, which were required to recall the "what, where, and when" components of a foraging and feeding event. This is the first behavioral evidence of such complex memory in an invertebrate.

SPATIAL MEMORY OF OCTOPUSES & CUTTLEFISHES

Laboratory and field experiments show clearly that octopuses can learn from spatial cues in their environment. Two impressive field studies—one in the Atlantic and one in the Pacific—show that octopuses can forage many feet during several hours amidst complex habitats such as coral reefs and find their way back to their dens. Sometimes the octopuses even terminate their foraging at a distance and swim on a "bee line" back to their den, indicating a refined sort of visual map that so far defies explanation. In a Pacific study of *Octopus cyanea*, it was particularly noteworthy that individual octopuses were foraging twice a day, covering an average of 330 ft (100 m) per day, while traversing four distinctly different coral reef habitat types and not foraging in the same areas on successive trips or on successive days. Their navigational skills and apparent spatial memory seem exceptional.

Cuttlefishes in the laboratory demonstrate an ability for spatial learning when tested in a variety of mazes in which escape was the reinforcing "reward." In some cases such as a wall maze or alley maze they learned within a few trials. However, in T mazes or open-field mazes, they did not learn at all. Cuttlefishes do not depend on home shelters the way that octopuses do, so their needs for spatial learning will be different (and yet to be determined) as these laboratory experiments showed.

RAPID ADAPTIVE COLORATION

PERHAPS THE MOST REMARKABLE FEATURE OF cephalopods is their complex and beautiful skin. Many species can radically change their appearance within the time it takes to blink your eye (about 200 milliseconds). This instantaneous change in appearance is used for camouflage as well as communication. These remarkable transformations are possible because brain centers send signals to nerves that run throughout the skin to a variety of pigmented chromatophore organs and reflective cells called iridophores. Many cephalopods control coloration by producing bioluminescence in the dark.

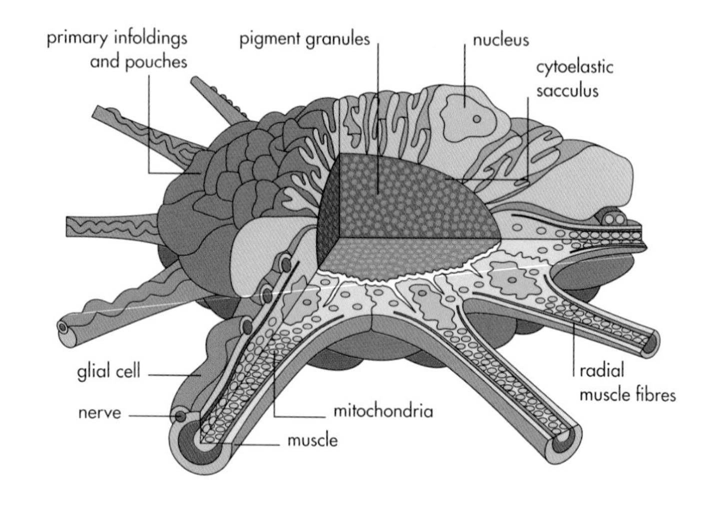

OPTICAL MAGIC: DYNAMIC SKIN PATTERNING

Chromatophores are like tiny elastic bags of pigment surrounded by a star-shaped arrangement of muscles. Because these muscles are under nervous control they can contract instantly to expand the pigment bag into a large disk. The color of the pigment may be red, brown, or yellow, depending on development and species. Each of the hundreds to thousands of chromatophores can be controlled individually by the cephalopod, allowing a wide variety of patterns to be either changed rapidly or held for prolonged periods.

Although some deep-sea species do not have functional chromatophores, this system is characteristic and unique to the neocoleoid cephalopods. On some areas of the body of many species the chromatophores overlie reflective tissue composed of iridophores. Light is reflected by layers of special cells called iridocytes. The animal can control color, contrast, brightness, and patterns on their bodies in less than one second by nervous control of the interaction of light with the pigments and reflectors in the skin. Some cephalopods also have permanent white patches, leucophores, just below the iridophores. The iridophores and leucophores can be hidden by expansion of the overlying chromatophores or exposed by their contraction, so producing patterns in their skin.

SHAPE-SHIFTING SKIN

In addition to the colorful body patterns, octopuses and cuttlefishes can morph their skin into three-dimensional shapes to enhance camouflage when nearby structures also have fine three-dimensional texture. These muscular "papillae" can also be expressed quickly and to any degree—from fully extended down to flat and invisible. Papillae come in a wide variety of shapes, from simply conical to bi-lobed, tri-lobed, or even flat. Some are short, some are long, some are fat, and others are skinny. Some papillae can be quite flamboyant in some octopods and cuttlefishes (see Flamboyant cuttlefish, page 96). The variety is impressive and they have not yet been cataloged and classified properly. No other animal on the planet can do this.

Some cephalopods have permanent skin texture that consists of various-shaped tubercles composed of cartilage or fibrous connective tissue. Distinctive papillae in some locations are sometimes referred to as cirri (superocular cirri, for example, are papillae above the eyes; in some species these are complex and may have embedded tubercles). Some octopods also have a semi-permanent lateral ridge of skin around the mantle.

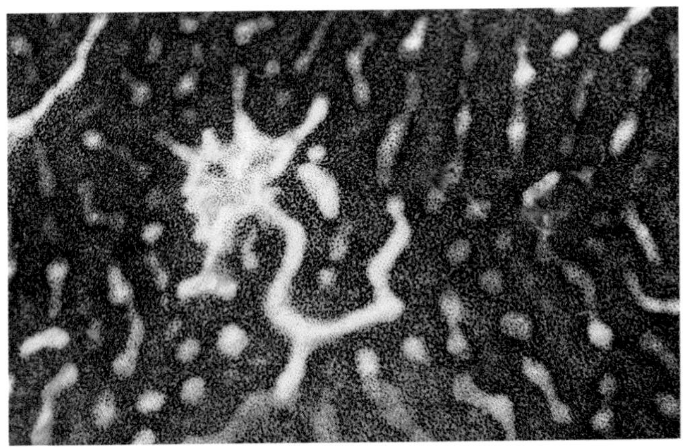

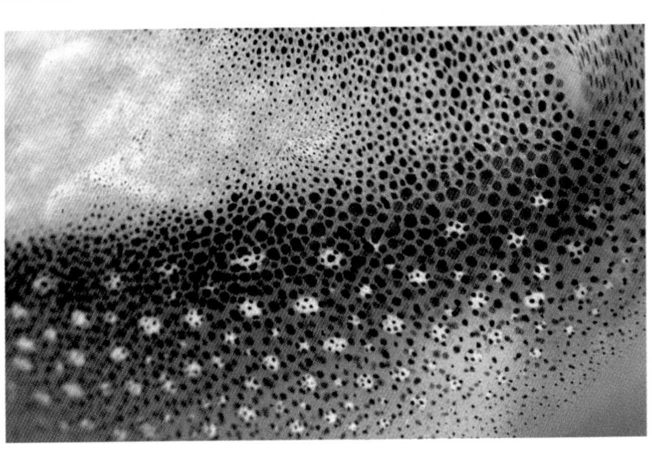

Opposite Diagram of the squid chromatophore organ illustrating the complexity of its many cell types that enable opening and closing in one fifth of a second to produce different skin patterns.

Left White leucophores in the skin of the California species *Octopus rubescens*; note the variable shapes and that some of them are obscured by expanded chromatophores.

Below left Close-up photograph of the skin of the cuttlefish *Sepia officinalis* showing the white leucophore spots surrounded by expanded red and brown chromatophores.

Below *Octopus tetricus* with extended skin papillae that produce a very textured skin to aid camouflage.

MAKING LIGHT WITH BIOLUMINESCENCE

Some cephalopods—mainly nocturnal or deep-sea species—have light-producing organs called photophores. There are two fundamentally different types. The first type, intrinsic photophores, produce light biochemically. The second type, the bacterial photophore, uses symbiotic light-producing bacteria that grow in special chambers within the host. These chambers are generally associated with the ink sac. Light can be produced within the bacterial chamber or, in some circumstances, the photogenic bacteria may be expelled with discharged ink, creating a glowing cloud in the water. Bacterial photophores are found in some species of the families Loliginidae and Sepiolidae, which use bioluminescence for counterillumination, the opposite of well-known countershading. Intrinsic photophores are known in many oegopsid squids and in vampires and are thought to be used in signaling as well as camouflage.

Octopods were long thought not to have photophores, but females of some pelagic incirrates are now known to develop large photophores around their mouths when they approach maturity. Also, one species of cirrate octopod (*Stauroteuthis syrtensis*) has been shown to have photophores associated with its suckers. Their behavioral functions are unknown.

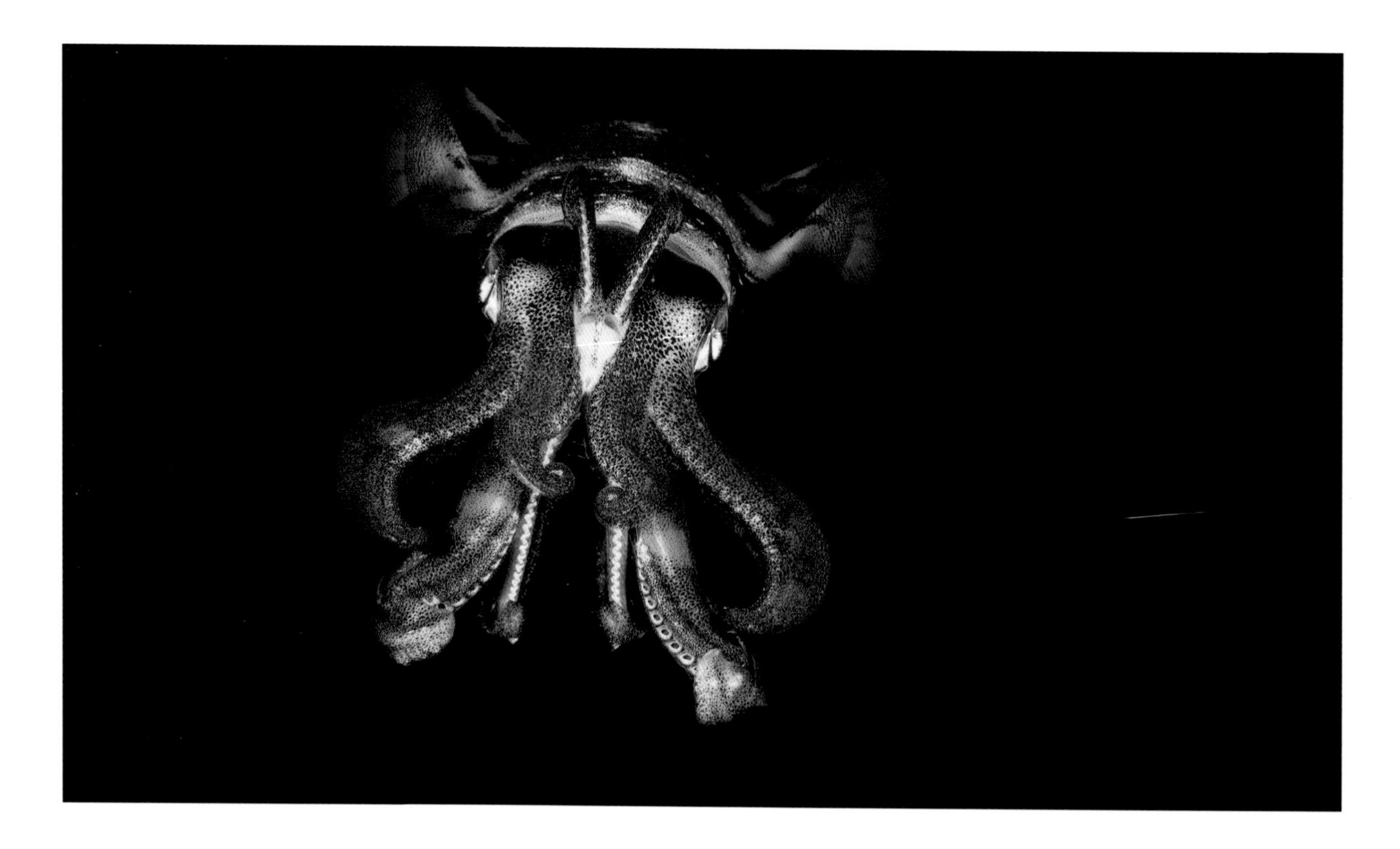

The light produced by both types of photophore is generally blue-green in color but this can be altered by structures associated with the photophore. As with the iridophores, flat platelets composed of the protein reflectin are formed to reflect and direct the light produced by the photophores. In some cases, the photophore lenses absorb selected wavelengths of the light, allowing other wavelengths to pass through. As a result, some photophores are blue, while others are green, and some are even red. In addition to structures that reflect and refract light, opaque structures around the photophores absorb light going in unwanted directions or at unwanted times. Some photophores have opaque lid-like structures that can be blinked open and shut to emit brief flashes of light.

Above The squid *Sepioteuthis lessoniana* at night showing a disruptive skin pattern. Note the broad fin around the mantle and the precise postures of the eight arms and two tentacles.

THE FAST COLORATION CHANGE PROCESS

It was recently established experimentally that cephalopods control their body patterning visually. For camouflage, they view their surroundings and quickly process that visual information, and then the central brain sends neural signals throughout the skin to millions of chromatophores and iridophores to produce the appropriate body pattern. Astonishingly this entire process can occur in as little as 200 milliseconds (one fifth of a second)! The diagram opposite illustrates this sophisticated process in cuttlefish.

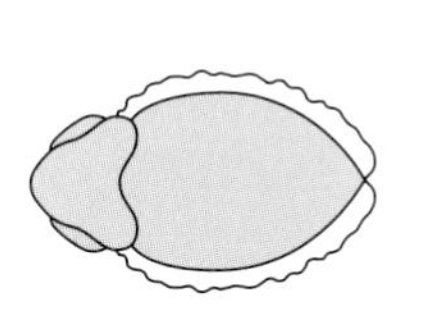
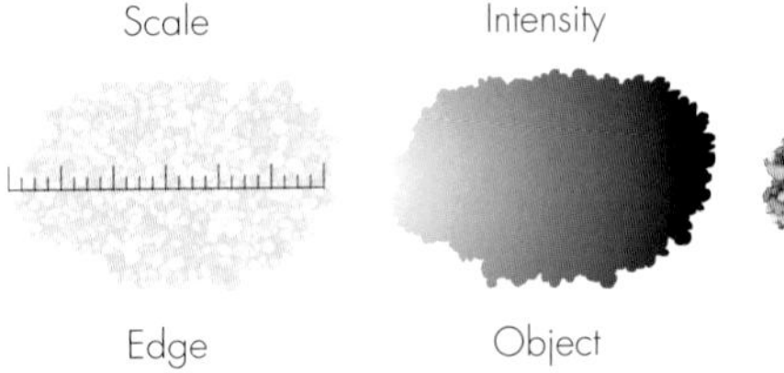

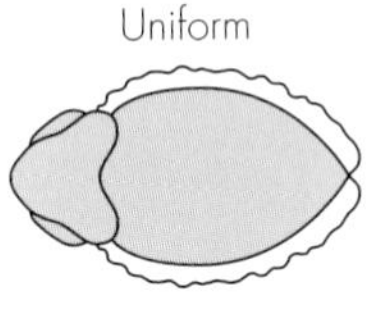
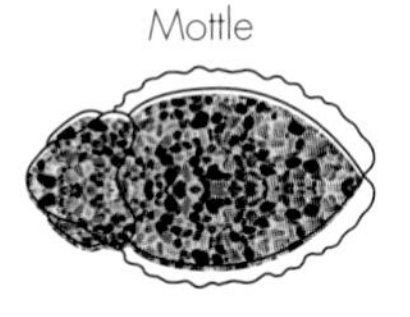
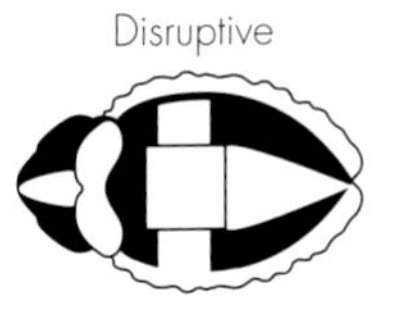
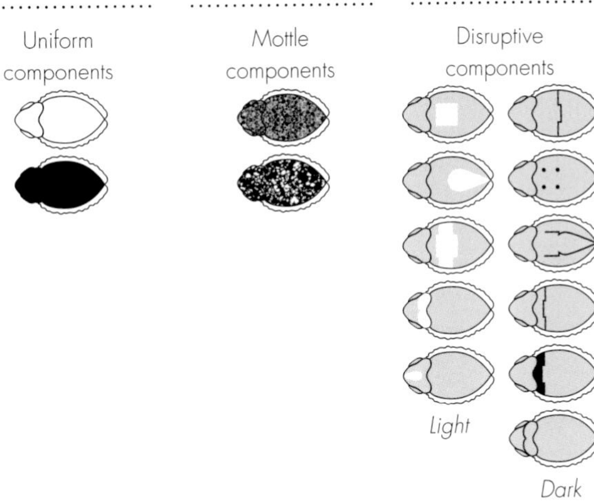
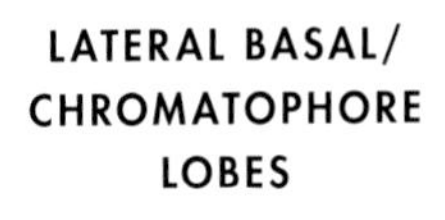

DECISION-MAKING
AVAILABLE CHOICES
NEURAL PATHWAY
SEES ITS ENVIRONMENT
Environment
Cuttlefish
EYE
ANALYZES ENVIRONMENT FOR CAMOUFLAGE
Scale
Intensity
Contrast
Edge
Object
Depth
OPTIC LOBE
SELECTS MODE OF CAMOUFLAGE
Uniform
Mottle
Disruptive
OPTIC LOBE
TAILORS PATTERN TO SPECIFIC SURROUNDINGS
Uniform components
Mottle components
Disruptive components
Light
Dark
LATERAL BASAL/ CHROMATOPHORE LOBES
ADAPTIVE COLORATION
End result
CHROMATOPHORE ORGANS

NATURE'S BEST CAMOUFLAGE

THE MASTERS OF CAMOUFLAGE ON THIS PLANET are universally accepted to be cephalopods, which can create optical illusions in the blink of an eye. An octopus or cuttlefish on a coral reef can produce effective camouflage on any of the hundreds of backgrounds that it encounters in this most complex of nature's habitats. The selective pressure for this unique capability emanates from the extraordinary variety of visual systems of multiple predators, particularly fishes, diving birds, and marine mammals.

Camouflage is all about visual perception. Primarily it centers around the visual capabilities of various predators, but in the case of cephalopods it is also about their own visual capabilities and decisions about which camouflage tactic to deploy. Thus, having the ability to change camouflage brings with it the burden of developing a method to swiftly analyze a visual scene, extract its key features, and then deploy an effective camouflage pattern. This is a remarkable cognitive process that involves decision-making at a high degree. It begs three basic questions: what are the fundamental tactics of camouflage that deceive such a wide range of predators? How many camouflage patterns does an individual cuttlefish or octopus have the ability to produce? How quickly can they change their camouflage?

DYNAMIC CHANGES TO AVOID DETECTION OR RECOGNITION

Octopus cyanea on Pacific coral reefs change their appearance more than 150 times per hour while foraging each day! Fast change is needed because cephalopods are mobile and move into different visual scenes constantly when foraging. During their stop-and-go foraging trek, they change their camouflage pattern each time they stop and it is tailored to that particular background. The speed of change is remarkable. By studying video of naturally foraging octopuses and cuttlefishes, it is known that they change in a fraction of a second. That is, they can create patterns in their skin in as little as 200 milliseconds; complex patterns may take up to 2,000 milliseconds (2 seconds) but usually it is in the order of 200–300 milliseconds, which is the speed of a human eyeblink.

Camouflage is something that cephalopods do almost all of the time because they are soft bodied and their primary defense is to not be detected or recognized in the first place. This presents different challenges when they are swimming in the water column or down near the bottom, where there can be a great deal of structure and color, or there can be open sand or mud plains that create different survival challenges.

The fundamental tactics are: 1) remain undetected; 2) disrupt body form so that the cephalopod is not recognizable as a distinctive octopus, cuttlefish, or squid; or 3) look like an uninteresting or distasteful object. Most animals in other phyla use one of these tricks, but cephalopods use all three with variations on each for unparalleled diversity.

How many camouflage patterns does a cuttlefish have? The European cuttlefish *Sepia officinalis* has been studied in the greatest detail both in its natural habitats and in the laboratory. Those extensive studies have revealed an unexpected finding: this species appears to have just three basic camouflage pattern templates (each with variations on the theme) to utilize on the wide range of backgrounds that it encounters over its very broad range. These patterns—called Uniform, Mottle, and Disruptive—are illustrated above right.

BACKGROUND MATCHING & DISRUPTIVE COLORATION

Resembling or matching the visual background is an effective method of remaining undetected and cephalopods excel at this. Uniform and Mottle patterns are shown by a wide variety of cephalopods such as the common European cuttlefish, which can swiftly change from conspicuous to Uniform or Mottle camouflage.

Left The three basic pattern types that cuttlefish use for camouflage: Uniform, Mottle, and Disruptive. Each individual can show these patterns.

Below This octopus is resembling the dark sand in pattern, color, brightness, and even the texture of its skin. The small white leucophore spots also represent a random sample of other small white pebbles in the background.

Above left *Sepia officinalis* in a conspicuous zebra pattern while swimming, then note the rapid change to camouflage in the next image.

Above The cuttlefish settled and in a fraction of a second deployed a Mottle pattern to match the background.

Disruptive patterning is a less obvious way to attain camouflage. Some disruptive patterns of cuttlefishes tend to blend into the background while other disruptive patterns—even shown by the same animal—are clearly detectable yet they render the cuttlefishes very difficult to recognize because they create false edges and put on light and dark patches of variable size, shape, and orientation so that it is hard to distinguish head from mantle, front from back.

Octopuses are very adept at background matching. However, due to their very flexible arms and non-rigid body mantle, they can contort themselves to cause visual disruption without creating as many skin patches as are needed by cuttlefishes or squids.

MIMICRY & MASQUERADE

Mimicry is looking like another animal, while masquerade is looking like an uninteresting object. Both tactics confuse recognition and both can be found in certain cephalopods. Mimicry is commonly understood to imply the resemblance of one animal (the mimic) to another animal (the model) such that a third animal (a predator) is deceived by their physical similarity into confusing the two. Only three cases of defensive mimicry are authenticated in cephalopods and all three are octopuses. The so-called "mimic octopus" *Thaumoctopus mimicus* (see page 176) and the undescribed "white V octopus" of tropical Indonesia, and the Caribbean long-arm octopus *Macrotritopus defilippi* mimic the shape, coloration, and swimming speed of local flatfish when they are moving. All three species live in open sandy flats where they would be detected by predators, so the guise is to look like some other animal.

Masquerade is widespread among shallow-water octopuses, cuttlefishes, and squids. To a human observer, cephalopods can sometimes look like stones, algae, seagrasses, and other objects such as soft and hard corals. Octopuses, cuttlefishes, and squids all use arm postures to enhance masquerade matching, and they are very good at using their vision to decide when and how to position their arms to help match nearby objects.

MOTION CAMOUFLAGE

Motion camouflage is moving in a fashion that decreases the probability of detection by a predator. Octopuses are good at this—they can move with stealth across open areas without being recognized. The cognitive aspects of this behavior are noteworthy because the speed of the octopus is generally similar to the speed of rippling light in that environment, suggesting that the octopus consciously regulates its stealth speed in accordance with ambient motion such as dappled sunlight from waves. Moreover, they can change their overall body shape and skin texture according to other three-dimensional objects in the distant background as shown in the illustration below, based on *Octopus vulgaris* in the Caribbean. That is, they can perform the "moving algae" trick when spiky algae are nearby, or the "moving rock" trick when smooth coral heads or rocks are nearby.

Above left When *Sepia officinalis* settle on pebbles or rocks with high-contrast light objects, they will respond with a Disruptive pattern to obscure their body outline.

Above Here the "mimic octopus" *Thaumoctopus mimicus* is mimicking a swimming flounder.

Below *Octopus vulgaris* can fine-tune its motion camouflage to resemble the shape of nearby algae (left), corals, or rocks (right) while it is moving.

WHEN CAMOUFLAGE FAILS

WHEN THE PRIMARY DEFENSE OF CAMOUFLAGE FAILS, as it sometimes does, cephalopods have a two-stage response of secondary defenses that startle then confuse predators. The first stage—termed deimatic behavior—is meant to make the predator hesitate in its attack sequence. The second stage—termed protean behavior—involves complexes of erratic unpredictable escape maneuvers, color changes, and behaviors. Together, these constitute a formidable defense strategy suitable for the soft-bodied cephalopods, which are otherwise quite defenseless.

MULTIPLE ESCAPE TRICKS

To interrupt the first stage of attack, cephalopods first use either a Stay or a Go tactic. The "Stay" tactic is a conspicuous deimatic display in which an octopus spreads the web between its arms to look bigger than it is, and makes a large dark spot around the eyes to startle the prey to make it hesitate just temporarily. Conversely they may immediately ink and jet away, which is a "Go" tactic.

Below "Fake right, go left" is the trick being used here to confuse a predator, whose eye is drawn to the black ink puffed to the right while the octopus blanches, quickly descends to the bottom and camouflages itself.

This is immediately followed by erratic unpredictable escape maneuvers. This phase of secondary defense is aptly called protean behavior after Proteus—the Greek god of unpredictability. These behaviors can be very complex—some octopuses can mix signals, maneuvers, and camouflage up to ten times in just 13 seconds. One such segment is the "fake right and go left" maneuver shown below. That one segment, extracted from video, took only 400 milliseconds.

TAILORED ESCAPE TO DIFFERENT PREDATORS

In filmed sequences underwater, as well as during laboratory experiments, it was found that cuttlefishes are capable of recognizing different predators and they modified their secondary defenses accordingly. This was true even for hatchling cuttlefishes that had no experience with predators, so this initial recognition represents an innate ability. Thereafter, however, the cuttlefishes were continually making decisions about how to implement protean behavior and they did so according to the microhabitat they were in at that time. This sort of swift sensory processing and decision-making is characteristic of high-level cognitive processing that large brains can enable.

Left This "deimatic display" of *Octopus vulgaris* is shown briefly to startle predators that have approached very close; the octopus is making itself look larger and different.

Below *Octopus vulgaris* performing the "Blanch—Ink—Jet Away" maneuver at high speed as secondary defense after its primary defense of camouflage fails.

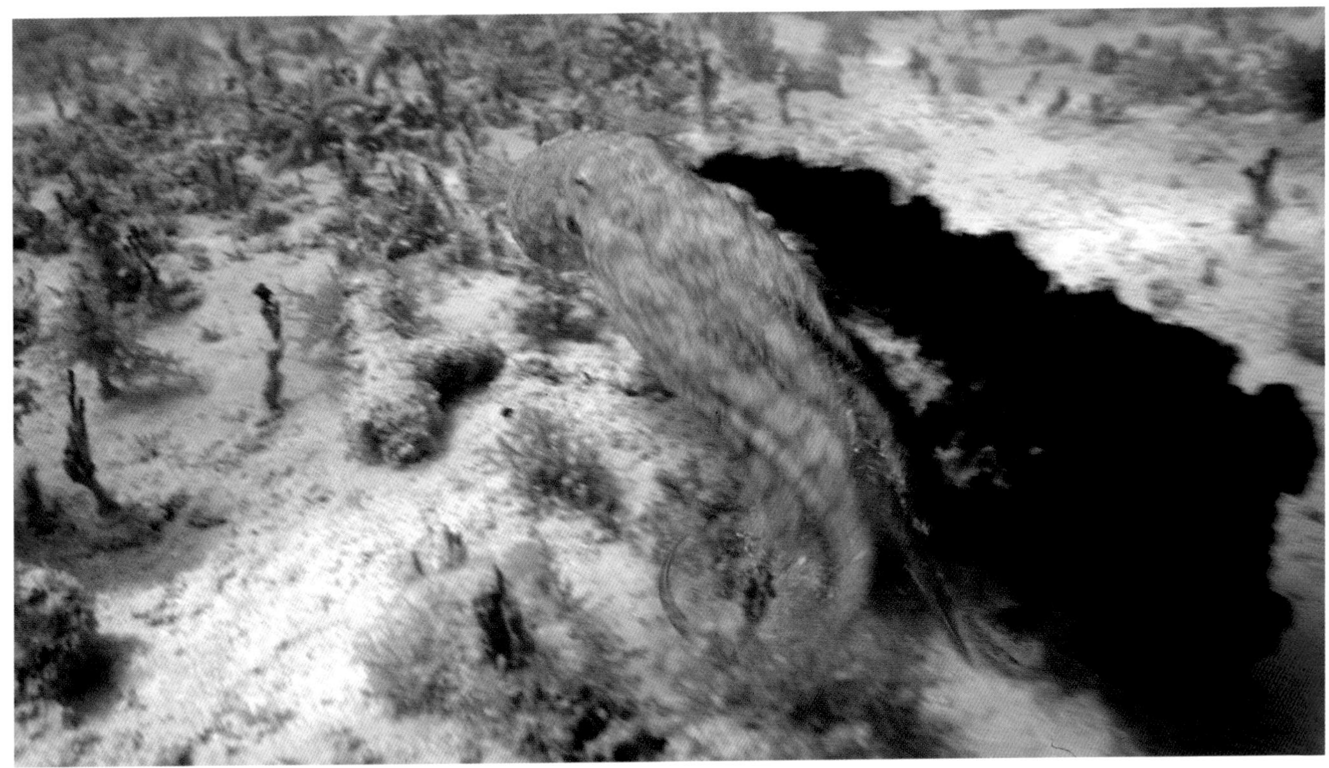

SUPER FIGHTS FOR MATES

BECAUSE MANY CEPHALOPODS LIVE FOR ONLY ONE year and reproduce once, males have forged a variety of tactics to acquire a female mate. The male–male aggression is intensive and there are long sequences of fighting bouts in which multiple visual signals and postures are used. Some of these displays rival those of birds and mammals, and some fights escalate to pushing and biting, but not to death. Contestants continually decide whether to withdraw or escalate fights depending on the opponent's actions.

ELABORATE GIANT AUSTRALIAN CUTTLEFISH DISPLAYS

Sepia apama in southern Australia aggregate once per year for spawning. Throughout the mass spawning season (May to August) there are about 180,000 cuttlefishes right at the shoreline in just a 2½ mile (4 km) stretch of rocky reef coastline. Some males are large, and others are small. Large males fight each other to gain some advantage in pairing temporarily with a female mate. Intense male–male aggression results in elaborate fighting displays, sometimes with roving dark bands moving over the body. Parts of these displays are very stereotyped and sometimes the males are locked in battle for many minutes. Recent experiments and field trials have shown that each male is continually assessing the other during these fights and adjusting his fighting tactic. This level of cognition compares with birds and mammals.

CARIBBEAN REEF SQUID DUETS

Sepioteuthis sepioidea is among the most sophisticated of squids and has both courtship and fighting displays. Females incite males with a bright white pied pattern that leads to male–male fights as elaborate as those of giant cuttlefishes. These fights are very short by comparison (a few seconds each) and the pattern on the top of the squid is different from that on the bottom of the squid, and each conveys a different level of aggression. Males can then show a silver pattern on one side to repel approaching males while showing the female a brown courtship pattern on the other side of its body.

Left Large male *Sepia apama* are fighting one another with flared arms, dark head and fins, and moving dark bands on the mantle to convey different levels of aggression.

Opposite top The female *Sepioteuthis sepioidea* (top) shows a white "pied pattern" that incites males to compete for her. The male (bottom) has begun to pair with her and will fight other males as they approach.

Opposite bottom The male (left) is showing a courtship pattern to the female while showing an aggressive "Unilateral Silver" on the other side to ward off approaching male rivals. When the female switches side the male immediately switches his pattern so that she does not see the aggressive pattern.

SNEAKY MALES & SNEAKIER FEMALES

SEX GAMES IN CEPHALOPODS ARE NOT A ONE-WAY street. For males, there are significant challenges to obtaining a mate because in some systems there are more males than females at the time and place of reproduction. Strong male competition has resulted in various sneaker tactics; that is, methods to obtain a mate that do not require fighting another male. For females, the challenge is to choose among many male suitors, and they can do this "cryptically" after mating whether the males are there or not.

CLEVER MALE TACTICS TO ACQUIRE A MATE

The ratio of males to females in the mating aggregation of *Sepia apama* in South Australia ranges from 4:1 to as high as 11:1, creating a strong bias that forces strong competition among males. Large males fight each other to become consorts to females, but smaller males are numerous and have only one spawning season to mate. They use fast dashes towards the female to obtain a quick copulation; when four or five small males are doing this all the time they distract the large consort male who cannot fend all of them off. If this does not work, small males will surreptitiously sneak under the rock where the female is laying her eggs and mate her out of sight of the large guarding consort male.

In the most extraordinary tactic of all, small *Sepia apama* males will "morph" into a female by hiding their fourth arms, which in males are very long and elaborate (see page 180). They then bulge their arms at the base as if holding a large egg as females do, and with this "sexual mimicry" they swim right past the large male, who thinks he is acquiring another female mate, and the sneaker male can then mate with the female.

Small males seem "pretty smart" because they are making many decisions during these active episodes. In fact, they are very flexible in their behaviors and are aware when the behavioral context changes. For example, if the large consort males leave for some reason, a small sneaker will move right up to a female and he will now act as a consort, not a sneaker, and he will fight off other sneakers. Squids can do this too. Such conditional mating strategies are characteristic of vertebrates including fishes and mammals.

On the other hand, the large consort male seems "pretty dumb" in these encounters because we know cuttlefishes have short-term and long-term memories, and he has just been chasing away that sneaker male for tens of minutes, and suddenly it morphs into a female and waltzes right in without the large male figuring it out. Not only that, but in a few cases the large consort male also tries to mate with the female mimic. Hence the visual deception on behalf of the small sneaker is highly effective.

CRYPTIC FEMALE CHOICE

Female cuttlefishes are also deciding with whom to mate. They reject 70 percent of mating attempts overall but sneakers gain many copulations, so sneaking is a successful male tactic, although it does not necessarily mean they gain fertilizations. Females on breeding grounds often have stored spermatophores from five to eight males and so have multiple sources of sperm available when they fertilize and lay each large egg. Large consort males that are guarding females as they lay eggs do not interfere with the female. This is the stage at which females are sneakier than males, because DNA fingerprinting shows that females are probably manipulating sperm in some unknown manner and "cryptically" choosing some sperm sources over others. Males are ignorant of these processes!

Opposite top A typical mating pair of the giant Australian cuttlefish, *Sepia apama*. The female is on the left. These pairs are very temporary, rarely lasting more than a day and sometimes only an hour.

Opposite bottom Sexual mimicry by a small male *Sepia apama*. The large consort male is reaching upwards to attempt to mate a small male that is imitating a female by hiding his dimorphic fourth arms, bulging his arms as if holding an egg, and displaying the female signal of white rim around his mantle. The female is below the large consort male, who is guarding her.

EVALUATING INTELLIGENCE IN SUCH BIZARRE ANIMALS

POPULAR OPINION IS THAT CEPHALOPODS ARE SMART or intelligent, yet it is very difficult to assess such qualities given our human biases. We have to immerse ourselves in the sensory world of an octopus, squid, or cuttlefish before we can begin to analyze its range of abilities. This has not yet been accomplished in an unequivocal scientific manner. Also, some behaviors of some cephalopods appear to human observers to be quite stupid and inappropriate. Let us review this.

Cephalopods certainly process many features that we equate with smartness or intelligence in vertebrate animals with which we, as humans, are more familiar: think of birds, rodents, and dogs. Intelligence, in most people's understanding, involves combining pieces of evidence to reach non-obvious conclusions. Cephalopod learning and memory capabilities are substantial, and most importantly it is clear from field and laboratory work on cephalopods that their behavioral repertoire is very wide by any standard, and they are making decisions constantly for a wide variety of interactions with predators, prey, and members of their own species. They have keen well-developed senses, a large complex brain, and a vast network of nerves throughout their skin and other body organs. It has often been postulated that cephalopods live a difficult life in the behavior space dominated by vertebrates, and that these selective pressures have forged the ecologically successful cephalopods we know today.

EMBODIED COGNITION

It is difficult to judge cognitive sophistication in an animal group with such a bizarre body organization. After all, these animals have their head on their foot (*cephalo pod*), they have a brain built on a molluscan plan, and they live in marine environments—with which most humans are unfamiliar. Perhaps it is more appropriate to judge their performance relative to the specific brain and body organization that they possess. This concept of "embodied cognition" has gained recent acceptance among neuroscientists and behavioral ecologists because it draws attention to the adaptive qualities that the animals possess. Anyone who has spent a lot of time underwater immersed in the sensory world of an octopus and has observed its wide range of behaviors over long time spans would agree that these animals clearly possess a brand of what humans call "intelligence."

ADVANCED INVERTEBRATES

In the terrestrial world—with which we humans are familiar—intelligence or smartness is calculated partly on behaviors such as exploration, play, tool use, or even consciousness. Moreover, social interactions also play a large part in such human assessments. However, except for certain squid species, cephalopods are not social. Efforts to demonstrate play and tool use in octopuses are just beginning and the results are not yet convincing; more experimental and field studies with several species are needed. There are no data available on consciousness although speculation continues.

However—and this is perhaps the most striking fact to emerge from the last 20 years of research—cephalopods are "advanced" invertebrates: almost every system that has been studied recently is extraordinarily sophisticated and unique, the result of over 200 million years of evolution. Finally, there is the intriguing possibility that cephalopods have evolved complex behaviors with cortical structure and neural pathways that differ from the vertebrate lineage. This idea deserves attention given that cephalopods are the only group to have diverged from the vertebrate line and evolved large nervous systems and complex behaviors.

Opposite This octopod is swimming in the water column, which is very rare for these bottom-dwelling creatures.

COMMON OCTOPUS

Octopus vulgaris

FAMILY	Octopodidae
OTHER NAMES	None
TYPICAL HABITAT	Seagrass beds, shallow coral reefs, rock reefs
SIZE	Mantle length to 8 in (20.5 cm)
FEEDING HABITS	Forages daily for mussels and small crabs; rarely eats fish
KEY BEHAVIORS	Day-active, sophisticated camouflage, spatial memory to find its way back to den

THE MOST COSMOPOLITAN SHALLOW-WATER octopus in the Atlantic is a master of camouflage in habitats as diverse as coral reefs, coral rubble, seagrass beds, rock reefs, and sometimes open mud and sand plains. It is found on both sides of the Atlantic from tropical to temperate, throughout the Mediterranean and Caribbean Seas, and is fished commercially off the Sahara Bank of Africa.

A COGNITIVELY ADVANCED CREATURE

The large complex brain of this octopus enables it to carry out a wide range of behaviors in complex habitats with many predators. Learning and memory are particularly well established in these bottom-dwelling cephalopods. This and other species conduct daily forages spanning dozens of feet and different microhabitats, yet they always find their way back to the den, presumably by spatial memory. Their soft pliable body enables them to squeeze into small holes or dens for shelter during most of the day and night. They use keen vision to forage on likely food spots and then feed in a tactile manner by enveloping coral heads and rocks that harbor shellfish and crabs. Some scientists think octopuses are capable of tool use and play although this is controversial.

STARTLING BEHAVIOR

This conspicuous deimatic display requires the octopus to "stand" on four of its eight arms and to spread its flexible arm webs maximally to look larger. It then produces dark high-contrast markings that make the eye appear larger as well. The octopus maneuvers so that it is always presenting its largest apparent size to the threat (usually a predator).

DAY OCTOPUS

Octopus cyanea

THIS LARGE, DAY-ACTIVE OCTOPUS IS VERY COMMON on coral reefs throughout the tropical Pacific Ocean. This is one of the largest octopus species and is among the most beautiful and colorful. Its camouflage is superb—it has to be since Pacific coral reefs have such a huge diversity of corals, sponges, and algae and there are dozens of rapacious predators looking for this high-protein meal. This graceful octopus is well known to divers and is also the target of small artisanal fisheries (mainly through spear fishing) throughout many of the Indo-Pacific islands.

KEYSTONE REEF PREDATOR

Octopus cyanea lives in large dens on the reef top and forages several hundred feet per day in search of mussels, clams, and crabs hiding among the corals and crevices. When these octopuses forage, small groupers follow them and feed on other small shrimps and fishes that escape as the octopus "parachute attacks" small rock promontories. Hence the octopus is grazing the filter-feeding molluscs and indirectly feeding small groupers. At the same time, the larger predators on the reef such as snappers and barracuda are preying upon the octopuses, particularly when they are small. The octopus is in the middle of the food chain on a coral reef and this key ecological niche that they occupy deserves conservation efforts to protect this keystone reef species.

FAMILY	Octopodidae
OTHER NAMES	Big blue octopus, Cyane's octopus
TYPICAL HABITAT	Shallow coral reefs throughout Indo-Pacific Oceans
SIZE	Mantle length 8–10 in (20–25 cm)
FEEDING HABITS	Preys on live crabs, mussels, and fish
KEY BEHAVIORS	Day-active, sophisticated camouflage ability, parachute attacks on likely food spots

KING OF CAMOUFLAGE

The day octopus may represent the pinnacle of sophistication among octopuses because it lives in the most complex of habitats: Pacific coral reefs with great structural diversity and biodiversity. During a typical 4-hour forage they may change their skin patterning more than 150 times to accomplish camouflage amidst different corals.

MIMIC OCTOPUS

Thaumoctopus mimicus

THIS SPECIES IS FROM A CLADE OF LONG-ARMED octopuses that have adapted to the bizarre underwater seascapes of open sand. These habitats are known to marine biologists as "muck dive sites" because they are devoid of coral or other structures, and are mostly characterized by seemingly featureless sand or mud bottoms. A shrimp, eel, fish, or cephalopod found in these habitats always has some specialized behavior enabling it to thrive in this strange environment.

ADAPTIVE IMITATION OF DIFFERENT ANIMALS

Both the Pacific and Atlantic mimic octopus species imitate swimming flounders and soles most of the time when they need to move swiftly and swim longer distances. The Indo-Pacific species has been reported to imitate up to a dozen animals such as shrimp, jellyfish, and snake eels, but there are few data to support these observations. Clearly this species can contort its body to look conspicuously like a variety of objects, many of which appear to a human to be other local species. Only recently discovered, it is not clear who the common predators of the mimic octopus are. Moreover, it would be informative to determine if these behaviors are learned or innate, and whether they are different in different locations across the wide range of these octopuses.

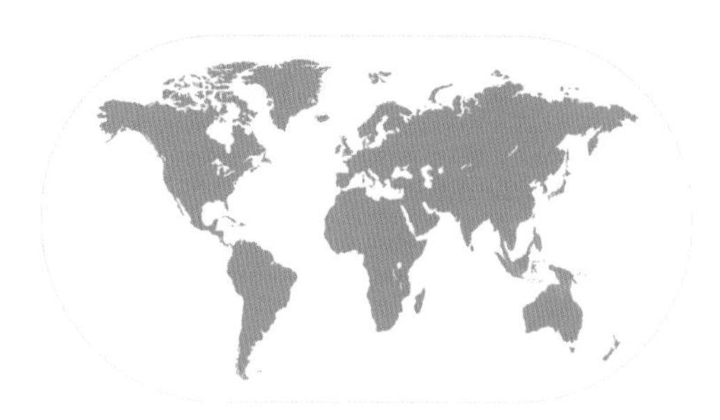

FAMILY	Octopodidae
OTHER NAMES	None
TYPICAL HABITAT	Sand and mud adjacent to coral reefs
SIZE	Mantle length about 2 in (5 cm), total length including the very long arms about 24 in (60 cm)
FEEDING HABITS	Preys on live crabs and shrimps
KEY BEHAVIORS	Aggressive imitation of a variety of marine species

SHAPE SHIFTER

The Indo-Pacific mimic octopus lives on open sand plains between coral reefs. In addition to camouflage for defense, it often mimics flounders and other sand-dwelling animals to confuse some of its predators while it is foraging to find food in these open flat habitats that do not provide hiding places.

COMMON EUROPEAN CUTTLEFISH

Sepia officinalis

FAMILY	Sepiidae
OTHER NAMES	Common cuttlefish
TYPICAL HABITAT	Sandy seagrass beds, shallow rock reefs
SIZE	Mantle length about 12 in (30 cm)
FEEDING HABITS	Lie and wait predator of small fishes; eats large crabs on occasion; hunts day and night
KEY BEHAVIORS	Superb camouflage, sometimes buries in sand; aggressive attack with two long tentacles

THIS FABLED SPECIES WAS FIRST WRITTEN ABOUT BY Aristotle, who marveled at its color-changing capabilities and the clouds of ink it emitted when alarmed. It has a wide geographic distribution in the eastern Atlantic Ocean from the North Sea southward throughout the Mediterranean and all the way to equatorial West Africa.

A WILY PREDATOR

Along with *Octopus vulgaris*, *Sepia officinalis* is one of the best-studied cephalopods in biology because it occurs in reasonable abundance near many marine laboratories throughout Europe. This cuttlefish is easy to keep or even culture in the laboratory. Hence, it has been very useful in elucidating how vision controls changeable camouflage, because any visual background you present it with will be quickly analyzed and answered with an appropriate body pattern. In nature, its camouflage capabilities are excellent, not only to deter predators, but to hide from unwary shrimps and fishes that it stalks slowly then captures with its lightning-fast strike using its two long tentacles. *Sepia officinalis* can also partially bury in the sand and act as a "lie and wait" predator. This species is fished commercially on a large scale on the Sahara Bank of northwest Africa.

RAPID COLOR-CHANGER
This "Zebra display" is typically shown by large males during intensive fights with other males. Each zebra skin pattern is unique, like a fingerprint. This pattern and others can be switched off or on in less than one-third of a second so that they can conduct rapid behavioral sequences with visual impact.

GIANT AUSTRALIAN CUTTLEFISH

Sepia apama

THIS NATIVE OF SOUTHERN AUSTRALIA GROWS UP to 26 lb (12 kg) and is the largest known cuttlefish species. Males have the long floppy and colorful fourth arm that is used as a sexual signal of male-ness. At the beginning of each austral winter (May/June), approximately 200,000 cuttlefishes migrate to a small spawning ground in northern Spencer Gulf, north of Adelaide.

ONLY CUTTLEFISH THAT FORMS HUGE SPAWNING AGGREGATIONS

This unique mass spawning is a "cuttlefish wonder of the world" and SCUBA divers can immediately see hundreds of cuttlefishes during a typical dive right next to shore. There are four to eleven males for every female on the spawning grounds, and this has led to some remarkable and sophisticated tactics by males to obtain a mate, including sneaker males and even sexual mimicry. Males conduct elaborate fights with one another to gain short-term consortships with females, who lay eggs under flat rocks as the males guard them against rivals. When the cuttlefishes are not competing for mates, they are camouflaging themselves in this rocky reef habitat. This is perhaps the best-studied and most complex mating system known in cephalopods and it rivals those of many vertebrate animals, including mammals.

FAMILY	Sepiidae
OTHER NAMES	Giant cuttlefish
TYPICAL HABITAT	Rock reefs, open sandy and seagrass beds
SIZE	Mantle length up to 20 in (51 cm)
FEEDING HABITS	Forages daily for small crabs and fishes
KEY BEHAVIORS	Solitary most of life cycle; day-active, sophisticated camouflage, complex reproductive behaviors

DISTINCTIVE MALE ARM
This large male cuttlefish is showing the elaborate green iridescence on its fourth right arm, which is very large compared with the other three arms on that side. This is the only visible dimorphic character that distinguishes males from females. Males also use this arm in fighting displays against other males when they are competing for female mates.

BROADCLUB CUTTLEFISH

Sepia latimanus

FAMILY	Sepiidae
OTHER NAMES	None
TYPICAL HABITAT	Coral reefs throughout the western Pacific
SIZE	Mantle length about 12 in (30 cm)
FEEDING HABITS	Forages daily for small fishes and large crabs
KEY BEHAVIORS	Solitary most of life cycle; day-active, sophisticated camouflage, complex reproductive behaviors

RAPID PATTERN CHANGES
The broadclub often deploys a characteristic white bar pattern (top) that disrupts its recognizable shape to predators. It can instantly change its body pattern to a uniform coloration (bottom) to help it resemble light-colored backgrounds such as coral or sand.

THE BROADCLUB OVERLAPS WITH THE DAY OCTOPUS on Pacific reefs and is day-active in these complex environments. This large and colorful species ranges widely in the western Pacific Ocean and lives exclusively on coral reefs, where it usually hovers motionless and undetectable like a mysterious flying saucer as it scans the scene for predators and prey.

CLEVER QUEEN OF PACIFIC REEFS

Despite their relatively large size and the clear water of coral reefs, these cuttlefishes are hard to find because they use so many camouflage tactics. While *Octopus cyanea* is the king of camouflage for reef bottom dwellers, *Sepia latimanus* might be considered the queen of camouflage on the same reefs: it does not sit on the bottom but hovers in and around corals and other reef structures. With its superb vision it determines where to forage in a stealth mode (that is, slowly and with changeable camouflage) and stalks small fishes. Occasionally it encounters a large crab, and then produces a dazzling visual display known as the "Passing cloud" in which fast-moving dark bands traverse its body—seeming to mesmerize the crab visually, which it then attacks. The cuttlefishes and octopuses can cohabit the same reef because they use different vertical space, forage differently, and eat different prey.

LONG-FINNED INSHORE SQUID

Doryteuthis pealeii

THIS SQUID LIVES IN LARGE SCHOOLS FOR MUCH OF its short life span of 6–12 months and it supports a large and valuable fishery off the eastern USA. It is a sleek predator capable of very fast jet propulsion (both backward and forward). Off the coast of New England, these squids migrate offshore to deeper water for the winter, then return to shallow nearshore spawning grounds in the summer.

THE SQUID NERVOUS SYSTEM STUDIED FOR NEUROSCIENCE

The long-finned inshore squid schools generally disperse at night to feed and aggregate in schools as a defense against predators during the day. Their large eyes are capable of seeing polarized light, which enables them to see and consume transparent prey organisms. When a large predatory fish approaches a school of squids, some of them will swiftly descend to the bottom and camouflage themselves by producing dark transverse bars across their body. Many aspects of the biology of this species are known because it has been studied for so long in Woods Hole, Massachusetts, where scientists from around the world gather during the summer. The nervous system, especially the giant axon, is a famous biomedical model.

FAMILY	Loliginidae
OTHER NAMES	Summer squid, Woods Hole squid
TYPICAL HABITAT	In summer, open sandy and seagrass beds near shore; in winter, in deep canyons offshore
SIZE	Adult males about 12 in (30 cm) long, females a bit smaller
FEEDING HABITS	Forages nightly for fish; sometimes cannibalistic
KEY BEHAVIORS	Schooling near the bottom; dispersing at night to feed over large areas; complex reproductive behaviors

MOVING IN FOR THE KILL
This sleek nearshore squid can jet very fast backward to avoid predators or forward to capture prey. The two long tentacles can just be seen in this photograph at the tip of the eight arms as the squid is about to attack a shrimp.

CARIBBEAN REEF SQUID

Sepioteuthis sepioidea

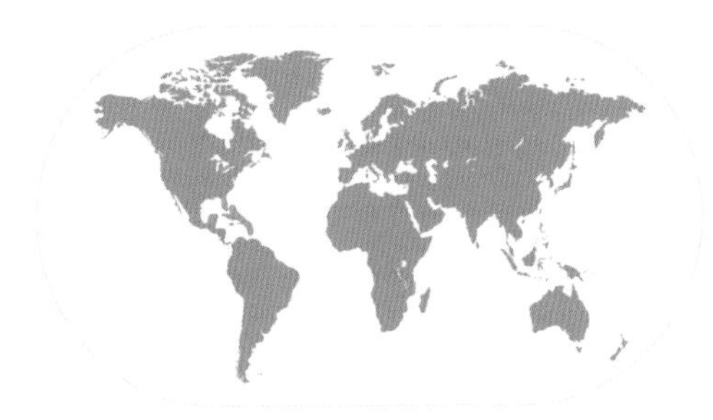

FAMILY	Loliginidae
OTHER NAMES	Reef squid
TYPICAL HABITAT	Shallow coral reefs
SIZE	Total length, including arms, to 8 in (20 cm)
FEEDING HABITS	Preys on small live fishes mainly at night
KEY BEHAVIORS	Huge signal repertoire, schooling, elaborate mating system

SQUIDS ARE SCHOOLING ANIMALS THAT HAVE social interactions, particularly with respect to sexual selection. The Caribbean reef squid has an elaborate mating system in which females show body patterns that incite the males to begin competing for them. The males then engage in action-packed fights among themselves using extravagant visual displays. Females are choosy and reject many male suitors. When a male pairs temporarily with a female, he will sometimes show a courtship pattern on the side of his body towards the female, and a fighting pattern on the other side of his body where rival males approach. This level of cognition is unknown among other invertebrates and compares to the sophisticated mating systems of fishes, birds, and mammals.

COLORFUL DISPLAYS FOR SEX AND THREAT

The reef squid lives in shallow, clear, bright water and produces the full rainbow of colors in its skin when signaling. Fighting and courtship patterns tend to be high-contrast to create unambiguous signals; examples are the "Pied" pattern of females and the "Zebra display" of males. When fish predators approach, the squids put on colorful patterns and postures that are tuned to the visual perception of those predators, for example, these squids display bright yellow in reaction to certain predators that approach closely.

GLITTERING COLORS

The Caribbean reef squid is one of the most beautifully colored cephalopods with skin that glitters from iridescence and pigments. They are commonly seen during the day in small schools comprising a wide size range of individuals. They have different body patterns for different behaviors.

CEPHALOPODS & HUMANS

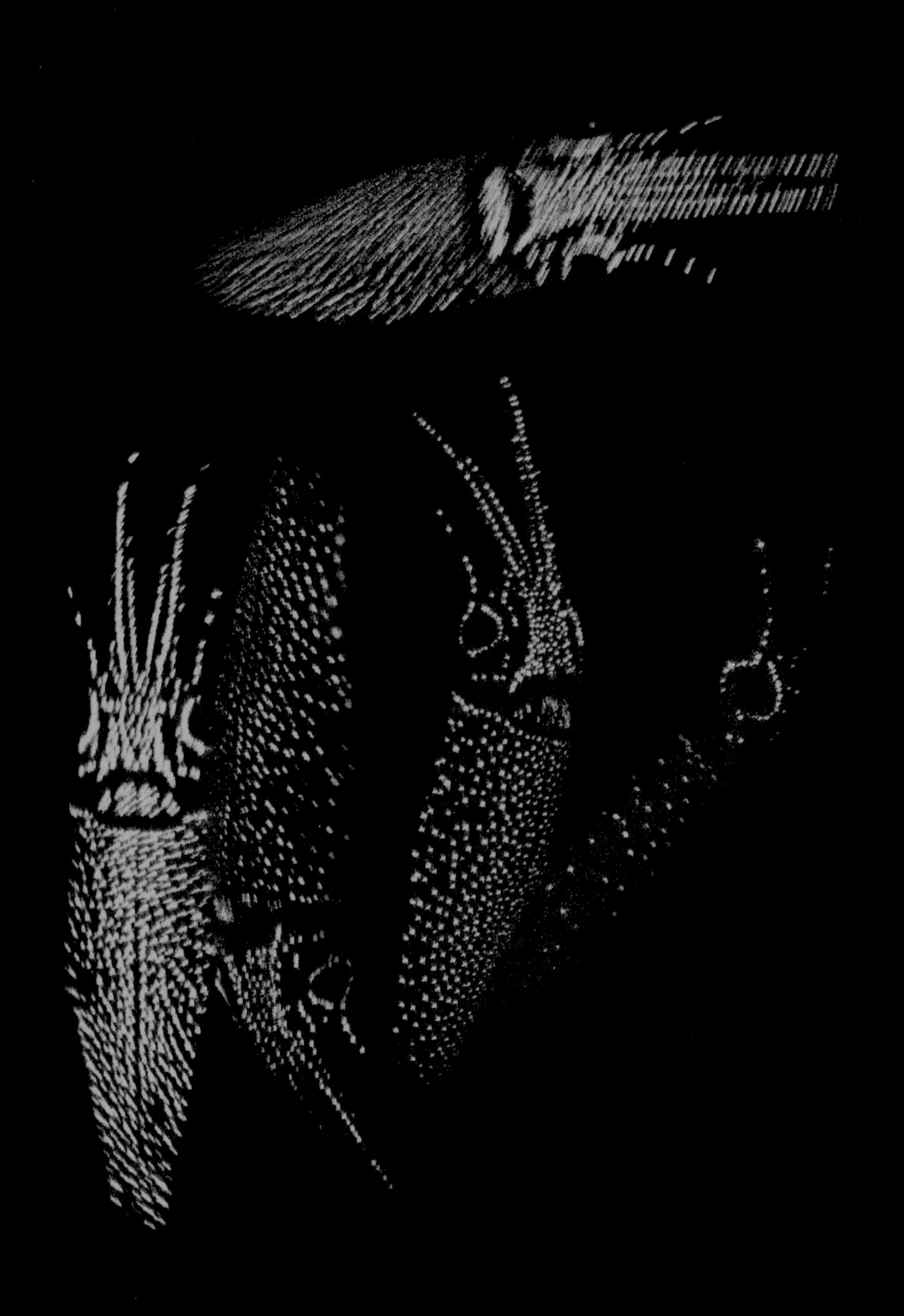

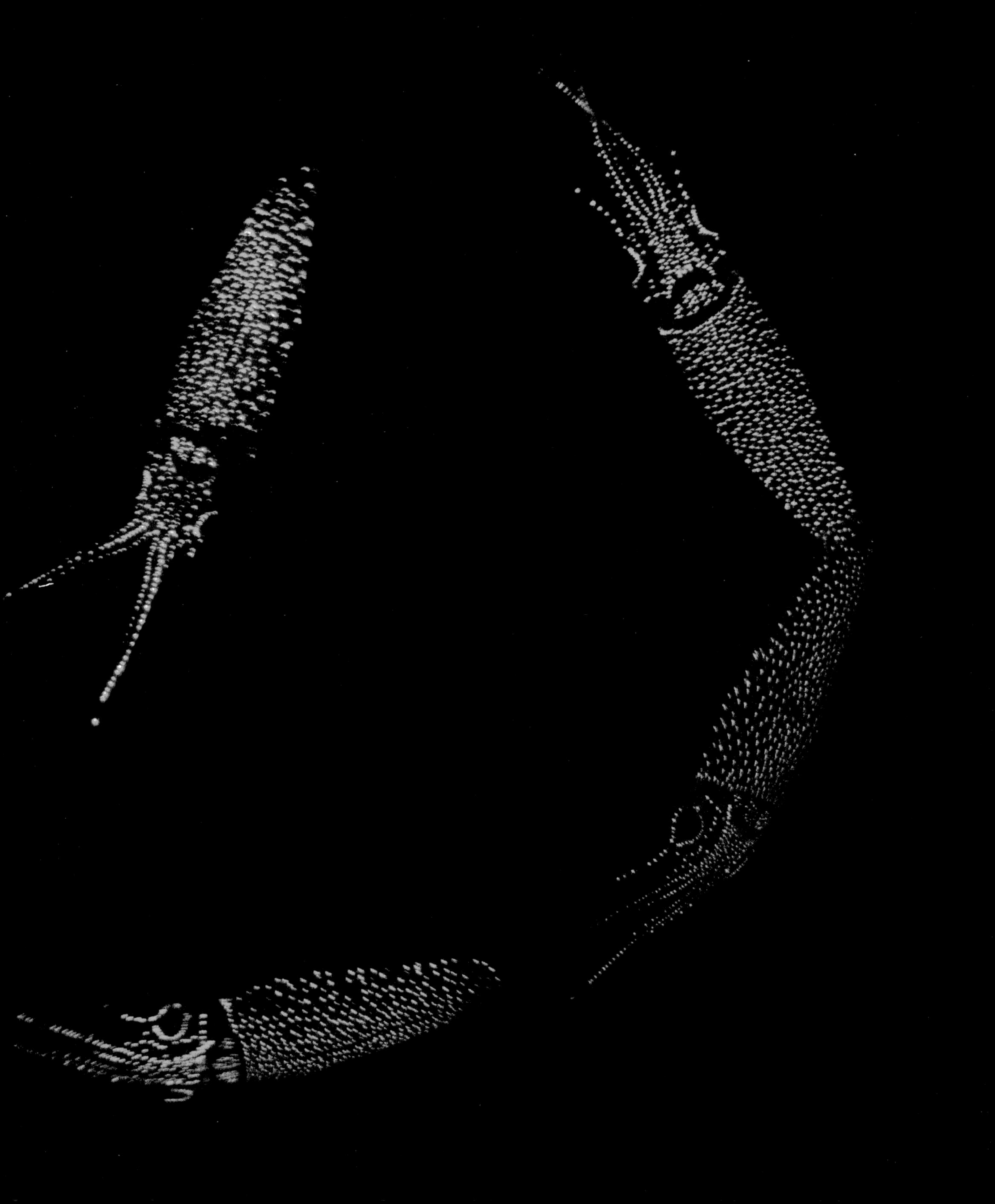

WORLD FISHERIES & HUMAN CONSUMPTION

CEPHALOPODS ARE AN EXCELLENT PROTEIN SOURCE, they are abundant in many parts of the world and when cooked properly some are very tasty. Consequently they have given rise to many commercial and artisanal fisheries throughout the world's oceans. From the biomedical perspective, cephalopods have made major contributions to understanding the nervous systems of many animals, including humans. The fields of materials science and engineering have also benefited by using cephalopod models as bio-inspired approaches to creating novel technologies that are useful throughout human society and industry.

The world catch of cephalopods (squid, octopus, and cuttlefish) increased from 3.4 to 4.7 million tons during the period of 1999–2015, reflecting an upward trend that had been occurring for several decades. This trend is partly attributed to the global shift from traditional to non-traditional target species, otherwise known as "fishing down the food web," when stocks of target species such as fishes become depleted. Cephalopods are prey and/or competitors to many traditional fish species. The total world capture figure has remained relatively steady in the past decade: in 2015 it was valued at 10 billion US dollars.

SQUID & CUTTLEFISH FISHING

Squid fisheries accounted for the majority of the global cephalopod catches: landings in 2015 were 3.5 million tons. Overall about 35 squid species have substantial commercial importance around the world but the majority of catch comes from just a few species. The oceanic ommastrephid squids such as the Humboldt squid *Dosidicus gigas*, and Argentine shortfin squid *Illex argentinus* dominate squid fisheries, each providing about 1 million tons annually. Japanese flying squid *Todarodes pacificus* landings are about 300,000 tons. These ommastrephids are captured at night mostly by using squid jigs (a type of lure with rows of barbless hooks at the bottom) and powerful incandescent lights that attract the squids to the vicinity of

the boat. This jig-fishing method exploits a key behavior of these aggressive predators: attraction to motion, which is indicative of live prey (squids are active carnivores). The automated jigging machines have elliptical drums that impart an up-and-down jigging motion that proves very efficient at hooking squids and pulling them swiftly to the boat.

Nearshore loliginid squid landings were more than 470,000 tons in 2015. *Doryteuthis gahi* (South American southern coastlines) and other species such as *Doryteuthis pealeii* (eastern coast of USA) are captured by trawling. *Doryteuthis opalescens* (off California) is strongly attracted to lights and captured by modified purse seines as squids aggregate under the lights, while the South African *Loligo reynaudii* is captured with squid jigs both at night with lights and during the day when they are actively spawning near the seafloor. Each commercial capture method is tailored to the schooling behavior, diel movements, and feeding methods of these highly mobile animals.

Opposite top Small-scale fisheries rely on the manual method of landing squids, using a handline.

Opposite bottom Typical squid jigs that look like fish or shrimp, both of which squid eat.

Above A Japanese squid jigging boat with a row of very bright lights and several rollers that are part of automated squid jiggers.

Above Opalescent inshore squid swarming at the surface after being attracted to the bright lights on the boat.

Opposite top Squid fishing boats in Vietnam at night, shining lights on the water surface to attract the squid.

Opposite bottom left Thickly aggregated squids, *Doryteuthis opalescens*, that have been attracted to the surface with bright lights. The large pipe vacuums them into the ship's hold.

Opposite bottom right Large marlin that has been hooked by a baited squid.

Cuttlefishes are benthic animals captured mainly by bottom trawling and a typical annual landing globally is 440,000 tons. Three quarters of this is taken in Asia, with around 150,000 tons landed in China, and an estimated 100,000 tons landed in India. More than 30 species have been identified as being of interest to fisheries. In Europe and Africa the majority of the catch is the common European cuttlefish *Sepia officinalis*.

THE OCTOPUS CATCH

Octopus fisheries landed 400,000 tons in 2015, of which 34,000 tons were the common octopus, *Octopus vulgaris*, on the Sahara Bank, the Atlantic coasts of southern Europe, and in the Mediterranean Sea. About 6,000 tons of *Eledone* species were landed in Europe and the Mediterranean, and 24,000 tons of *Octopus maya* were landed in Mexico. The rest was captured worldwide, mainly in Asia. Small-scale artisanal octopus fisheries exist all around the world. Asian countries dominate the market demand for cephalopods. However, restaurants all around the world—especially Europe and even North America—are increasing the menu choices with cephalopod recipes.

DEMAND FOR CEPHALOPODS

Cephalopods are also of use to humans in less noticeable ways. As they are in the middle of the food chain, they are a primary food for many recreational game fish such as marlin, grouper, snapper, and flounder. One example is the highly valuable sport-fishing industry for striped bass along the eastern seaboard of the USA, which uses live squid as bait.

Overall, it is expected that fishing pressure on cephalopods will increase in coming decades. Only a few stocks are at maximum exploitation, and as cephalopods have very short life cycles and most reproduce in large numbers, they have the flexible life history to respond to change relatively well. It is hoped that effective and conservative management practices can be implemented consistently in the future.

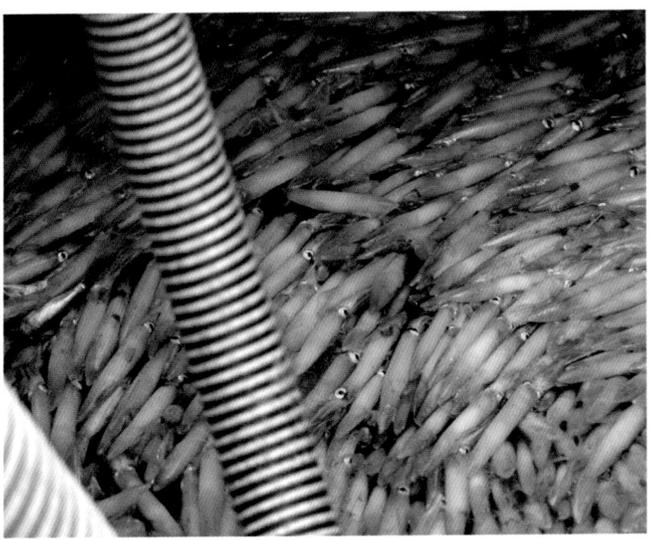

A RICH HISTORY OF BIOMEDICAL & BIOLOGICAL ADVANCES

THE 1963 NOBEL PRIZE IN PHYSIOLOGY OR Medicine was based largely on the squid giant axon, which provided a model to prove for the first time that nerve cells work by conducting an electrical signal. Every medical student and neuroscientist knows this famous preparation, which is still in use today. Many other cephalopod features have been developed for medical and biological research.

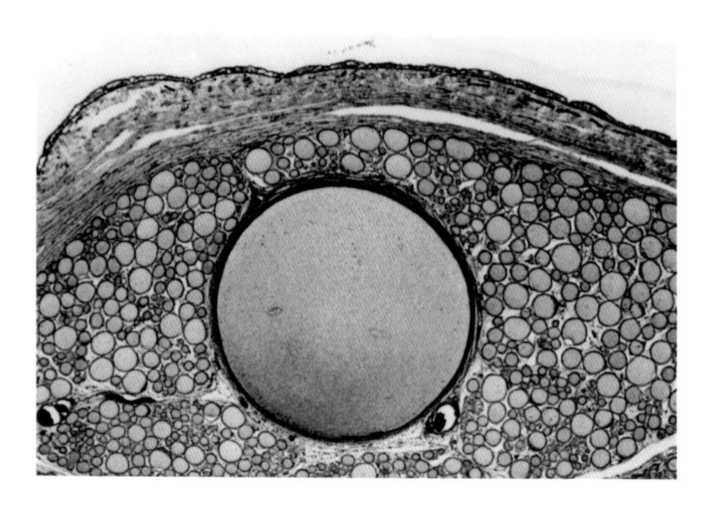

Above This very large axon, or nerve fiber, enables fast jet propulsion in squids. Compare its diameter with other surrounding axons.

Opposite The squid *Doryteuthis pealeii*, which is commonly used in Woods Hole, Massachusetts, for neuroscience studies of the giant axon.

THE SQUID GIANT AXON FOR NEUROSCIENCE

Neuroscience and human medicine have benefited greatly over many decades from the pioneering work performed with the squid giant axon as a model of a single neuron. This "giant axon" is indeed giant: its diameter is approximately 100 times that of the typical human nerve fiber. It was first mistaken for a blood vessel because it was so large. It is broad enough to insert electrodes inside and outside the cell membrane and to measure the complex of ion channels that make up the standard machinery for all nerve fibers, including humans'.

Discovered in the 1930s, this model nerve fiber was studied intensely during the 1950s and 1960s. Many hundreds of research papers have been published on this model preparation. The genome for the squid *Doryteuthis pealeii* is now being sequenced, and this will open up new avenues of research on what is perhaps the best-known nerve cell in all of biology and human medicine.

So why do squids have a giant axon? In many ways it is an evolutionary dead-end—a genuine oddity in the animal kingdom. Squids are capable of jet propulsion and fast escape; and to squeeze the mantle forcefully for maximum jetting power, the brain needs to transmit signals to the most distant part of its large mantle rapidly, so that all of the mantle contracts in unison. Rapid conduction of electrical signals is achieved by dramatically widening the diameter of the nerve fiber, which cuts down the internal resistance to the electrical signal. Vertebrate animals, including humans, have evolved an insulating myelin sheath (or Schwann cell) around nerve fibers that speeds conduction velocity, so that we do not need unusually wide nerve fibers to operate our bodies and we can fit many more rapidly conducting fibers into our nerves. Fortunately, human ingenuity has found ways to exploit the oddball squid axon to better understand how nerve cells work, or sometimes do not work, as in diseases such as Alzheimer's.

OCTOPUS MODEL FOR LEARNING & MEMORY

The octopus model for learning and memory is another famous preparation developed in the 1950s by the venerable Prof. J. Z. Young, who, incidentally, first developed and popularized the squid giant axon. He and his colleagues were some of the first to relate specific actions to individual brain lobes, including centers for learning and memory. For the general experimental paradigm, *Octopus vulgaris* that were trained in the Naples zoological station were able to learn visual discrimination on

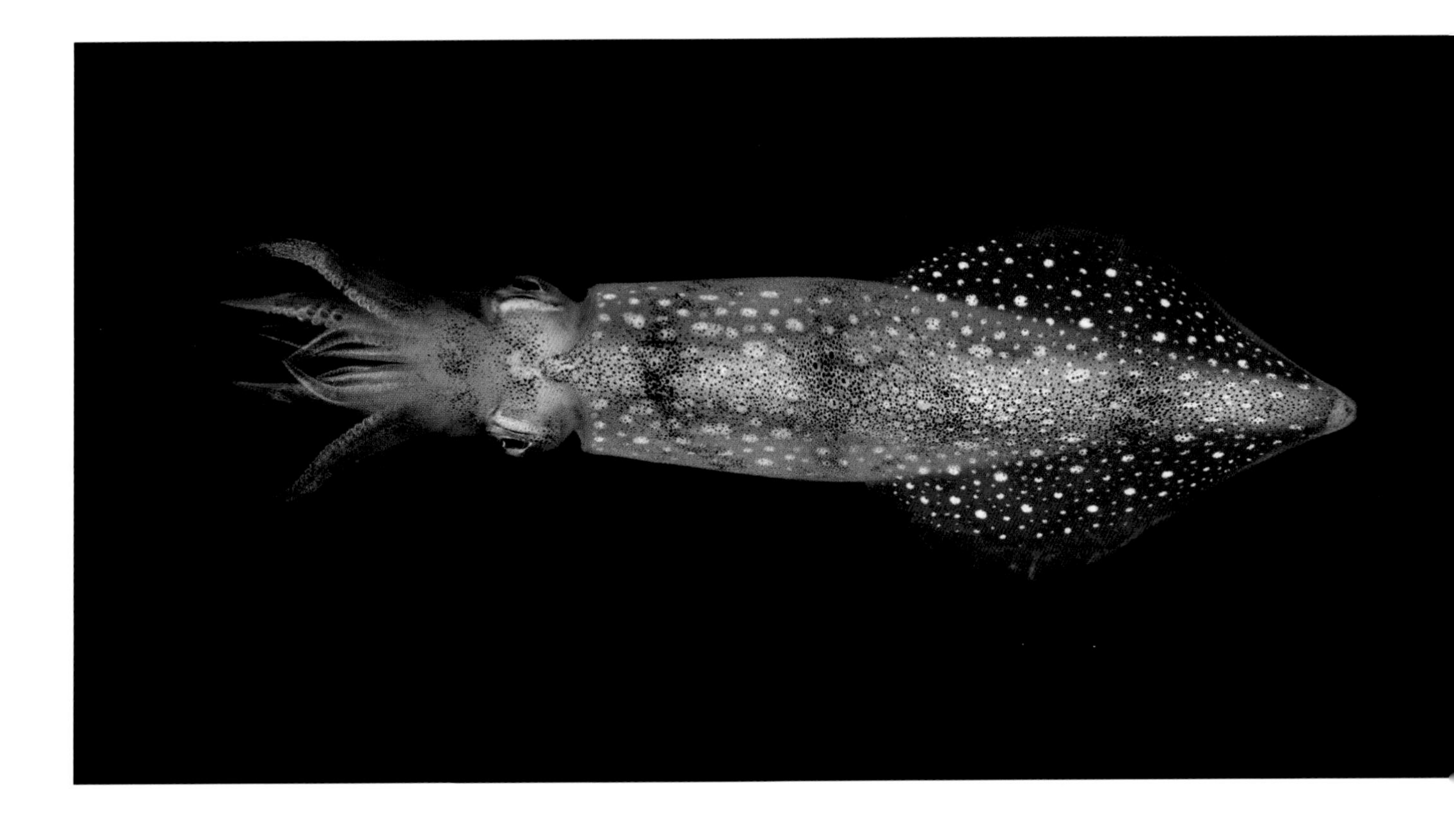

just the second or third day of training and were shown to produce both short-term and long-term memory, as in many vertebrate animals. Other British colleagues worked out the details of touch learning and its neural bases. These landmark studies stimulated large communities of neuroscientists to study the neural basis of learning and memory in simpler organisms, because the octopus brain was too large and had too many small cells to make a practical model at that time. Curiously, there is now renewed interest in cephalopod learning and memory, with exciting new investigations of spatial learning and even episodic memory, which was thought to be an attribute only of advanced vertebrate animals. There are new methods to reconsider studying cephalopod learning and memory at the cellular and molecular level, and exciting discoveries await us in the near future.

OCTOPUS & SQUID GENOMES

The first cephalopod genome to be sequenced and annotated was that of *Octopus bimaculoides* as reported in the journal *Nature* in 2015. Several remarkable findings were uncovered in this landmark publication. In particular, the genome size is enormous—nearly the size of mouse and human genomes! Hundreds of cephalopod-specific genes were discovered, especially those that showed elevated expression in the skin, the suckers, and the nervous system, all of which support the general consensus that these animals have complex and unique behaviors.

Recent corroborative research also shows that cephalopods edit their nervous system RNA at an astonishing rate to create a vast array of new proteins, more than in any known

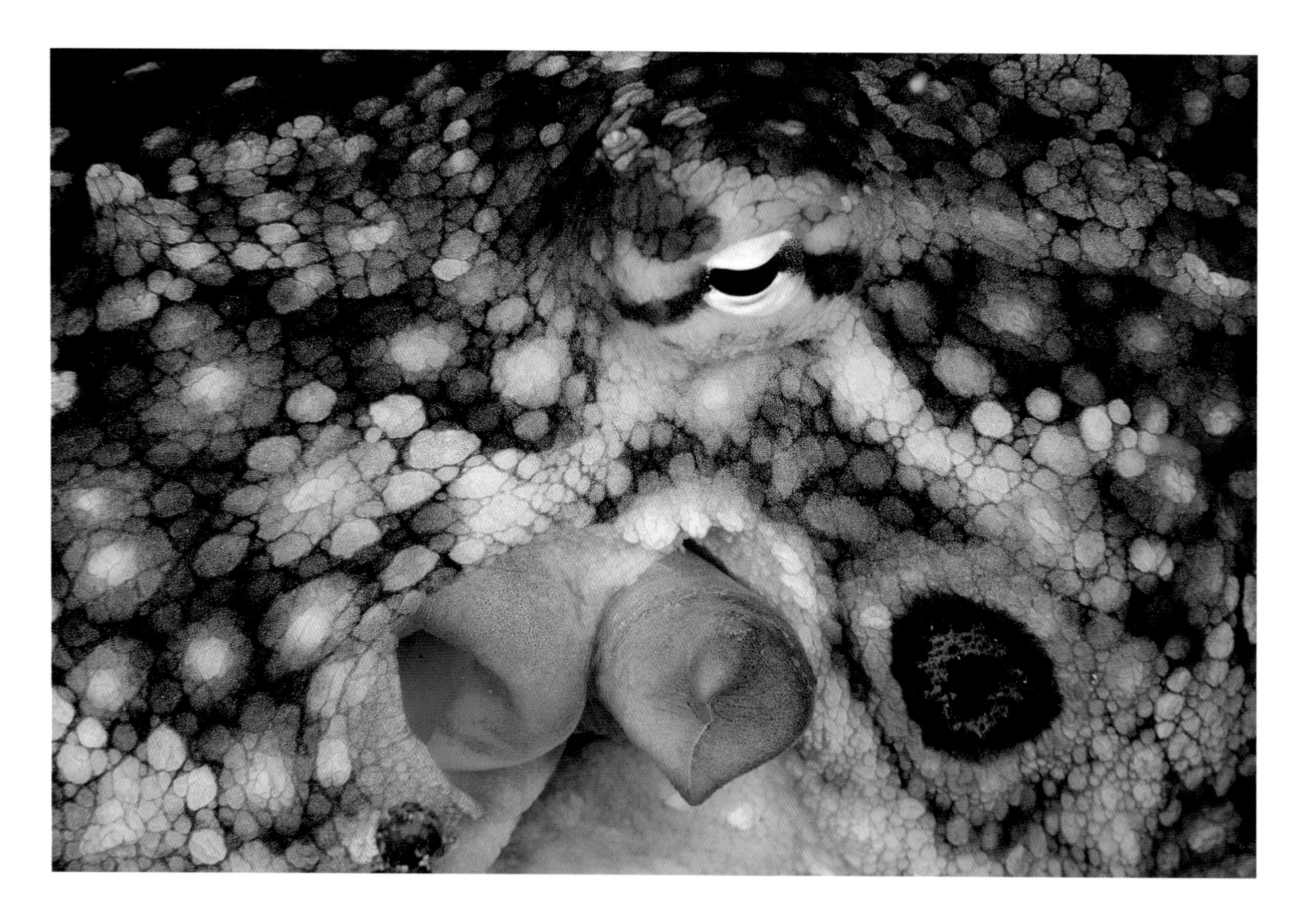

organism (including humans). By editing RNA, cephalopods gain the flexibility to express those proteins required under a specific set of environmental conditions. However, this comes at a cost: in order to edit their RNA, their rates of DNA-based genome evolution have slowed. The significance of these findings for neuroscience in general and for biomedical applications is now being studied. The genome for the squid *Doryteuthis pealeii* is due to be sequenced next, and this commercially and biomedically important species also has a huge genome, which was certainly not anticipated. Many aspects of biological study will be advanced with these new genomic data. Another dozen or so cephalopods are in the process of having their genomes sequenced and the next decade will be an exciting time for research.

BOBTAIL BIOLUMINESCENCE: A BIOMEDICAL MODEL OF SYMBIOSIS

The light organ of the Hawaiian bobtail creates light by recruiting new luminescent bacteria every day so that the bobtail can have vibrant bioluminescence for counter-illumination during the night when it forages. This process has become a major biomedical model of deciphering the

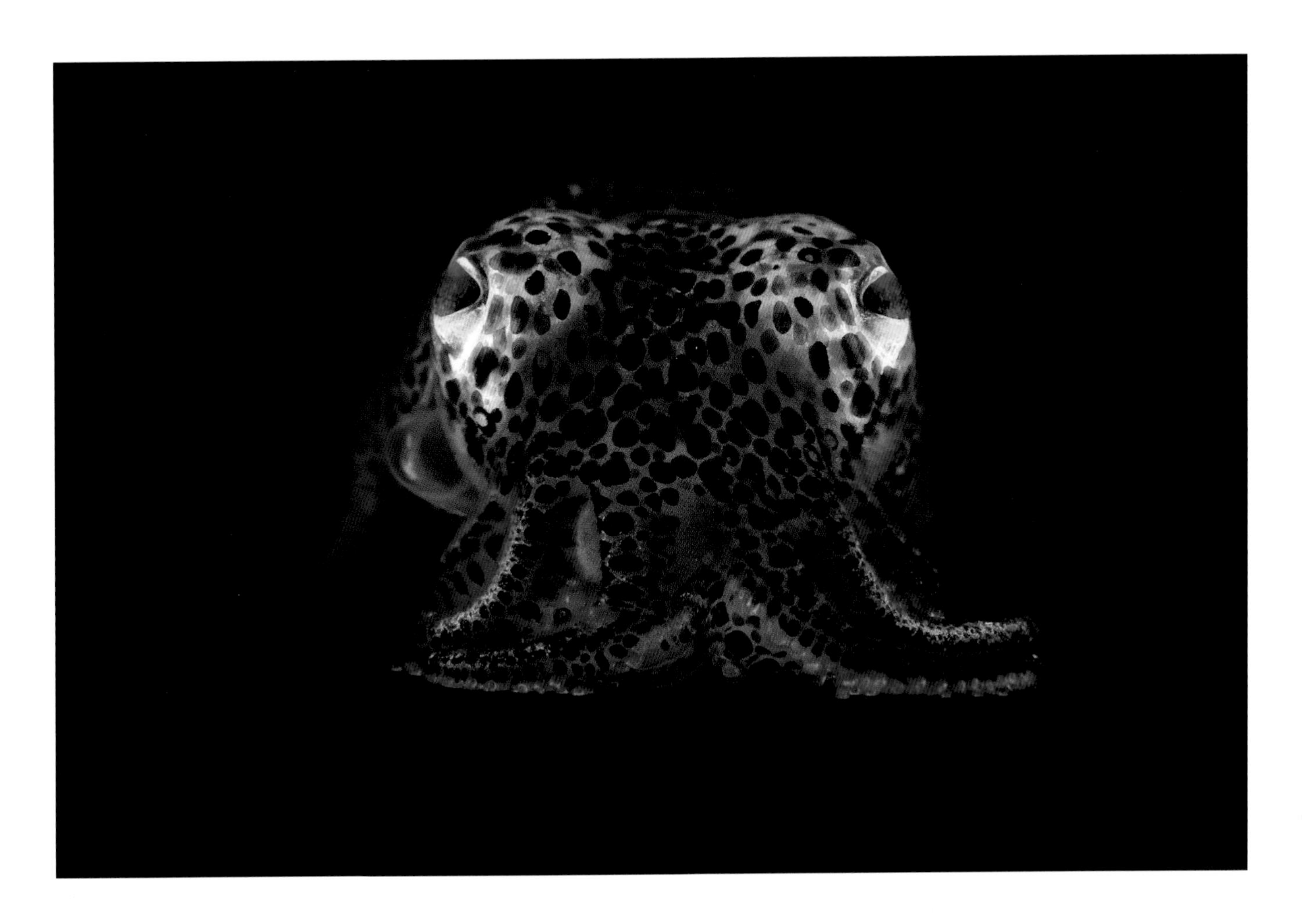

molecular conversation that takes place between the host tissue (the light organ) and the symbiont (the luminous *Vibrio fischeri*). Most vibrio bacteria are pathogenic to animals and humans (as in cholera, which is caused by *Vibrio cholerae*) but in the bobtail the vibrios form a healthy and normal symbiosis.

What are the fundamental processes that govern a healthy versus a pathogenic association between various animal tissues and environmental bacteria? This question has direct relevance to human health. The squid/bacterial model has uncovered many details that can help to answer it.

Opposite *Octopus bimaculoides* with its distinctive blue iridescent ocellus.

Above The Hawaian bobtail *Euprymna scolopes* with pigmented chromatophores expanded and blue iridescence expressed on its arms.

Within the wide realm of biology, various other cephalopod model organisms have contributed to our understanding of subjects as wide ranging as aggression, vision, and reproduction. For example, the cuttlefish has proved in the last decade to be an informative model to understand visual perception as it relates to rapid adaptive camouflage.

BIO-INSPIRED MATERIALS SCIENCE & ENGINEERING

Top The colorful *Octopus mototi* has complex skin for camouflage.

Above University of Illinois scientists made this flexible material that senses what is under it (in this case the university's acronym) in a compelling demonstration of how cephalopod skin biology can inspire new classes of engineered materials.

Opposite The green, blue, and silver colors in this squid's skin are reflections from iridophores, many of which are turned on and tuned to different colors by nerve fibers. Fabrication of artificial materials based on these biophotonic studies are underway.

CEPHALOPODS CAN DO MANY THINGS THAT humans would like to be able to do, such as change the color or pattern of their clothes, cars, or complexion. This futuristic dream is close to reality given the advances in technology as well as in-depth investigations of the biological processes that enable these capabilities in cephalopods. Even robots with flexible arms based on those of octopuses are being developed for a wide range of uses.

NEW CLASSES OF COLOR-CHANGING MATERIALS

The dynamically changing skin of cephalopods provides a biological model that may have widespread translational value to a variety of human needs and desires. The ways that humans could use this capability are matched only by the imagination. On the trivial side, changing the appearance of our clothing for a changing social scene could create fun and laughter. On a more serious note, hunters, soldiers, or police could make themselves more inconspicuous or conspicuous depending upon the nature of the endeavor. To enable the transition from basic to applied science we must understand how changeable colors and patterns are produced.

In cephalopods, the simplest explanation is that they use elegant combinations of pigments and reflectors in their skin to create the optical illusions for camouflage or communication. Recent discoveries have demonstrated how pigments and reflectors in cephalopods manipulate light at the level of the wavelength of light, thus paving the way for scientists to create novel classes of nanomaterials that could also change color and pattern.

Materials scientists and engineers have joined with biologists to create some prototype skins that are both flexible and changeable in appearance. One such example, shown on the left, is based upon flexible cephalopod skin, and this device is capable of producing black-and-white patterns that

spontaneously match those of the surroundings without user input or external measurement. This approach demonstrates a complete set of materials, components, fabrication methods, integration schemes, bio-inspired designs, and coordinated operational modes for adaptive optical electronic sheets.

The beautiful iridescence that is found in the skin of almost all cephalopods is produced by a novel protein found only in cephalopods, called reflectin. This remarkable protein can dynamically change its conformation to turn iridescence on or off, and can even tune its color across all of the rainbow's spectrum. Materials scientists have extracted this protein and, to their surprise, the protein molecules self-organized themselves into thin films and reflected light, paving the way to future optical devices including fiber optics.

The reflectin protein has also been engineered into thin films that can be dynamically tuned over more than 600 nanometers in bandwidth (humans can see across a 300-nanometer range, from 400 to 700 nanometers). This tunability allows the films to reversibly disappear and reappear when visualized with an infrared camera, so leading the way to controllable camouflage coatings on a variety of surfaces.

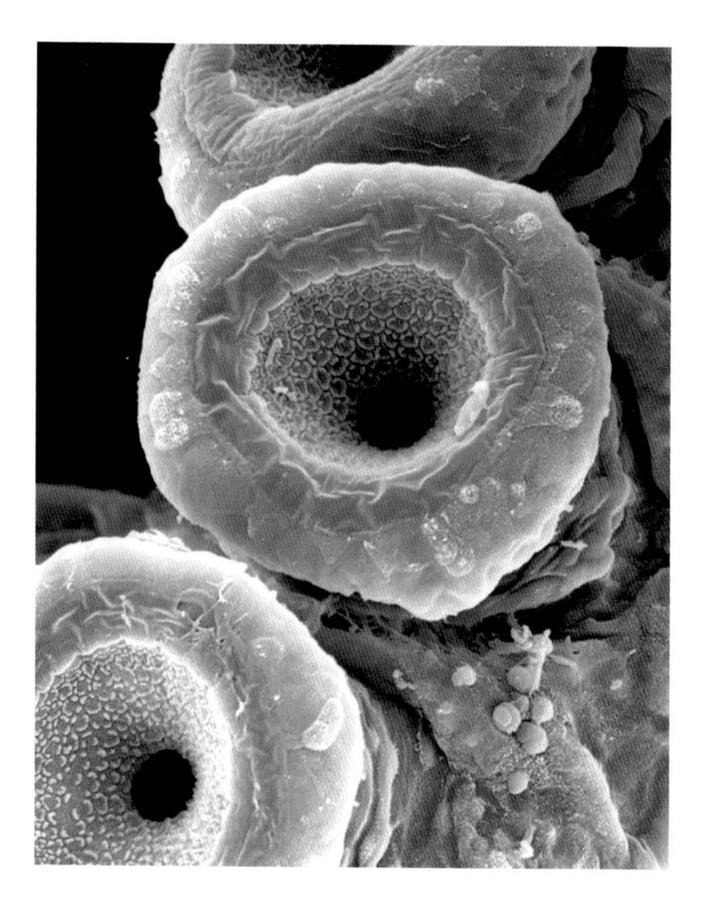

Above This scanning electron microscope image of octopus suckers shows some of the intracacy of their morphology. Each sucker also has thousands of sensory receptors for touch and taste.

Opposite These squid suckers have a toothed ring of tough yet flexible protein that passively "bites" into the skin or scales of prey items to cut down shear forces.

SOFT ROBOTICS

Robotic engineering has also profited from imitating the soft, flexible, yet strong arms of octopuses. Whole robots that can move in constrained and twisted environments (such as collapsed buildings) are in demand, and research groups in Italy have designed and built octopus-inspired robots that are gaining these capabilities.

A single robotic arm appendage with multiple degrees of bending, twisting, and extension is sought for medical applications, and prototypes of such devices are being tested. For example, a European team of collaborators has used the octopus arm as a model to develop what they call "Stiffness controllable flexible and learn-able manipulator for surgical operations" or STIFF-FLOP. Their approach is to combine artificial cognition and robotics to design and test more flexible devices that can move, deform, and change stiffness for minimally invasive surgery (via endoscopy) and other medical applications.

SUCKERS FOR FLEXIBLE ATTACHMENT DEVICES

Cephalopods have suckers on their arms and tentacles and there is substantial variety in their morphology and functionality. The octopus sucker is a complex organ that attaches forcefully to wet, smooth, or rough surfaces of great diversity. Engineers and biologists are measuring their structure and mechanical properties and beginning to build a new generation of attachment devices that are pliable to many terrestrial and manufactured substrate profiles whether wet or dry. These will be useful on robotic arms among other devices.

Squid suckers have an unusual feature that enhances their ability to hold wriggling slippery prey—that is, to counteract the shear forces applied during horizontal movement relative to the sucker surface. This "toothed ring" is very sharp and made of a tough protein that can bend and deform just enough to hold the prey but not detach from the sucker musculature. The porous architecture of the protein enables this combination of mechanical attributes that could find applications in industry and society. These are very challenging endeavors but the potential payoffs are considerable.

HORNED OCTOPUS

Eledone cirrhosa

FAMILY	Eledonidae
OTHER NAMES	Curled octopus (UK)
TYPICAL HABITAT	Sand, mud, broken rock, and rocky reefs 200–400 ft (60–120 m) depth on continental shelf
SIZE	Mantle length up to 7½ in (19 cm)
FEEDING HABITS	Wide range of prey including shrimps, lobsters, crabs, brittle stars, polychaetes, gastropods, and fishes
KEY BEHAVIORS	Typical octopus bottom dweller living in dens and using camouflage while foraging

SIMPLE SEA-BED DISGUISE

This is a typical bottom-dwelling octopus that eats a large array of live prey. Its body patterning is quite drab compared to octopods that live in shallower, well-lit and diverse habitats. This octopus has only one series of suckers down each arm, whereas most octopus species have two series.

HORNED OCTOPUSES ARE SOLITARY ANIMALS. They readily enter lobster pots and prey on lobsters, crabs, and other trapped species. North Sea populations of horned octopus increased recently as some fish predators of octopus were overfished, and this has a cascade effect of reducing the commercial crab and lobster fisheries. Horned octopuses are commercially fished with bottom trawls and they are harvested on a large scale in the Mediterranean Sea. It is curious that they can be caught in trawls since they live in dens most of the day. They may pick up the low-frequency vibrations of the oncoming bottom trawl and then respond by jetting upward out of their den into the water column, which places them into the mouth of the fast-moving trawl net.

OCTOPUS OF BIOLOGICAL VALUE

The physiology and morphology of skin patterning have been studied in the horned octopus as a comparison to other octopuses that have richer and more diverse skin and patterning repertoires. Due to its relatively simple skin, the motor fields of chromatophores on the mantle have been mapped by stimulating nerve centers in the brain. The white spots in the skin are called leucophores and are useful for camouflage as well as taxonomic identification. Its name refers to a large protruding papilla above each eye that resembles a horn.

LESSER TWO-SPOTTED OCTOPUS

Octopus bimaculoides

THIS SHALLOW-WATER ADAPTABLE OCTOPUS is common in southern California and it lives not only amidst beautiful kelp forests and rock reefs but on muddy bottoms close to shore, an indication of its adaptability to visually and structurally diverse habitats. It forms its dens in existing crevices and holes in most habitats, but when residing in the flat muddy habitats this species constructs elaborate burrows that are lined with rocks, indicating an ability to manipulate the local environment in a clever and adaptive fashion. This species forages daily to feed on a wide range of live prey including shelled molluscs, shrimps, and crabs. Its camouflage in these habitats is superb.

FIRST CEPHALOPOD GENOME TO BE SEQUENCED

The first fully sequenced cephalopod genome ever published (in the journal *Nature* in 2015) and made available to the wide scientific community was *Octopus bimaculoides*, and it was chosen as the most suitable candidate because (1) it is typical in body form and behavior of the many species of octopus, (2) it is common and available to the research community in the USA, and (3) this species has large eggs and is relatively easy to rear or culture in captivity. The genome is very large, which was not unexpected given the complexity of the nervous system and behavior of the octopus, and hundreds of cephalopod-specific and octopus-specific genes were discovered, many of which showed elevated expression levels in specialized structures such as the skin, suckers, and nervous system. This landmark study lays the groundwork for a great deal of exciting and novel future research in many scientific disciplines.

FAMILY	Octopodidae
OTHER NAMES	The California mudflat octopus
TYPICAL HABITAT	Temperate rock reefs and kelp habitats; muddy flats in shallow water
SIZE	Mantle length to about 4 in (10 cm)
FEEDING HABITS	Forages daily for bivalves, gastropods, shrimps, and crabs.
KEY BEHAVIORS	Day-active, sophisticated camouflage, sometimes constructs complex rock-lined burrows in mudflats

BLUE-RINGED EYESPOTS

The distinctive eyespots, or ocelli (one on each side of the octopus), are displayed when the octopus is startled or threatened. Otherwise the ocelli are indistinguishable from the coloration of the rest of the body. Only a few octopus species have ocelli.

SOUTHERN BLUE-RINGED OCTOPUS

Hapalochlaena maculosa

FAMILY	Octopodidae
OTHER NAMES	Lesser blue-ringed octopus
TYPICAL HABITAT	Seagrass beds, sand and rubble, shallow subtidal reefs
SIZE	Mantle length up to 2¼ in (57 mm)
FEEDING HABITS	Forages nightly for small crabs and shrimp
KEY BEHAVIORS	Superb camouflage, flashing blue rings when startled or threatened

WITH ITS IRIDESCENT FLASHING RINGS this octopus warns approaching predators of its toxicity. This is a small species and this defense tactic is quite unusual among cephalopods. The blue rings are hidden most of the time and in fact this is one of the best-camouflaged of octopuses, but when it is discovered by a predator it will change its appearance dramatically and this startles the predator and presumably makes it hesitate briefly so that the octopus can ink and jet away.

THE MOST VENOMOUS CEPHALOPOD

Not only does this octopus have venomous toxin in its salivary gland to paralyze large prey, but the potent neurotoxin tetrodotoxin (TTX) is also distributed in all body parts with high concentration in the arms followed by the abdomen and gills. A bite to a human can be lethal because the TTX produces respiratory arrest within minutes and immediate artificial respiration must be given for survival. Fortunately, this octopus is not aggressive and it requires a human to handle (or mishandle) the octopus before it will bite the person. So the simple solution is to heed the warning of the blue rings and not touch or manipulate this octopus species without great care. The mother invests TTX into her eggs so that the young are venomous upon hatching.

BLUE FOR DANGER

The threat display of flashing blue rings is highly conspicuous and is considered a warning display to deter close approach of predators. This display is honest because this octopus is extremely toxic to predators and to humans. The fast flashes are achieved by relaxation of the muscles above the iridescent rings; upon fast neural relaxation of those muscles the rings are suddenly exposed and appear bright blue within a second. Sometimes the iridescence is in rings or spots (top) or short stripes (bottom).

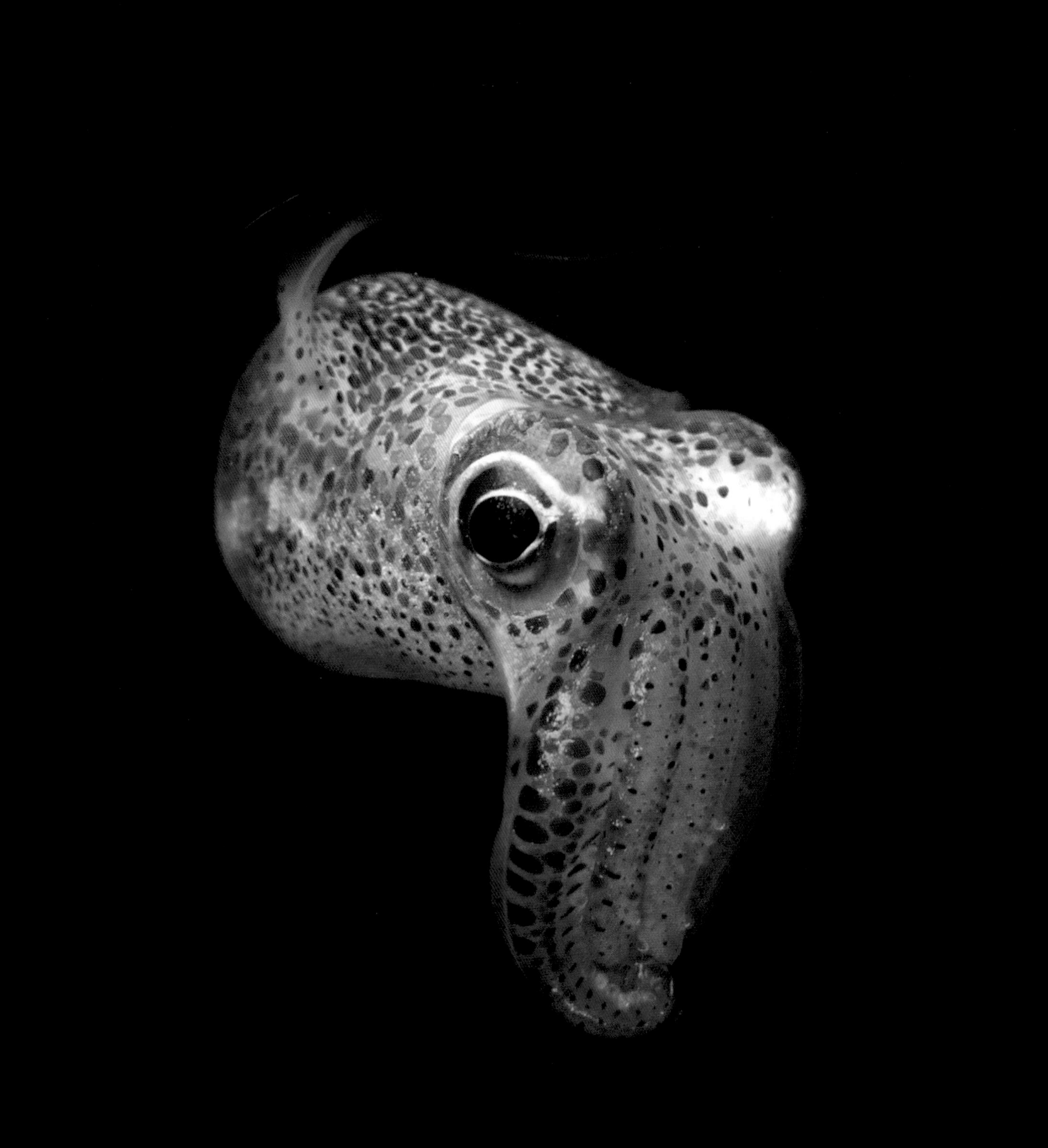

HAWAIIAN BOBTAIL

Euprymna scolopes

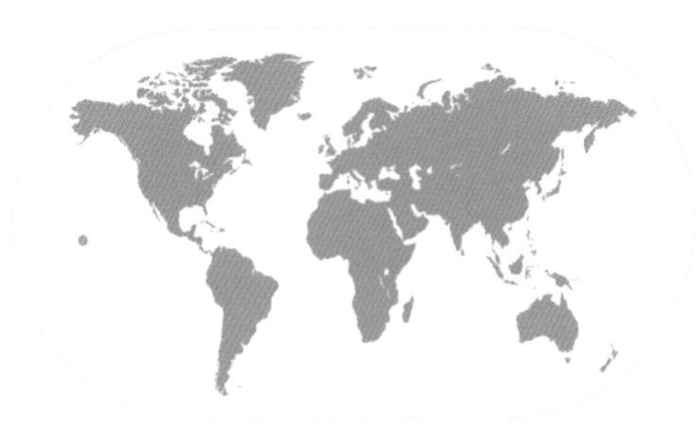

FAMILY	Sepiolidae
OTHER NAMES	Hawaiian bobtail squid, Mu he'e
TYPICAL HABITAT	Sandy substrates close to shore usually in shallow water
SIZE	Mantle length up to 1³⁄₁₆ in (3cm)
FEEDING HABITS	Mainly live shrimps during the night
KEY BEHAVIORS	Buries in sand during day; swims in water column at night using large bioluminescent light organ for counterillumination

THESE SMALL AGGRESSIVE SAND DWELLERS have two particularly clever adaptations to living on the open sand plains around Hawaii. First, they bury in the sand to hide during the day, with only their eyes protruding (but well camouflaged); they also have a glue that keeps the sand in place on their topside, and they can instantly throw the sandy glue layer off to rapidly jet away from a predator or grab a shrimp. Second, they have a very large light organ in their mantle that helps them camouflage (via counterillumination) at night while swimming in the water column.

A KEY BIOMEDICAL MODEL

Their luminescent organ produces light with the common seawater bacterium *Vibrio fischeri*. Each morning the bobtail expels the bacteria from its light organ and then recruits new bacteria from the seawater during the day to start a new culture of fresh bacteria to create luminescence that night. The intriguing aspect of this is that most bacteria of the genus *Vibrio* are pathogens to many animals including humans: in the bobtail the bacteria are symbiotic—not pathogenic. Comparing *Vibrio fischeri* with the pathogenic *Vibrio* helps medical researchers learn more about the molecular signals that enable symbiosis versus pathogenesis, which could lead eventually to better antibacterial medicines. The genome of *Euprymna scolopes* is recently completed and the bobtail is relatively easy to culture so this may become the first cephalopod lab model for continuous culture.

DAYTIME HIDING

This small bobtail is not technically a squid—it is in the order Sepiolida. It is fully adapted to a sand environment and hides in the sand all day then forages during the night using its bioluminescent organ for countershading camouflage. Note the large chromatophores and the iridescent splotches all over the body.

CAPE HOPE SQUID

Loligo reynaudii

FAMILY	Loliginidae
OTHER NAMES	The chokka squid
TYPICAL HABITAT	Continental shelf at 30–650 ft (10–200 m) depth; sometimes offshore down to 1,150 ft (350m)
SIZE	Mantle length up to 16 in (40 cm)
FEEDING HABITS	Paralarvae feed on copepods and adult squids feed mainly on fishes but also on crustaceans and polychaetes
KEY BEHAVIORS	Schooling near the substrate during the day, swimming in water column at night and feeding. Complex reproductive behaviors

LARGE MATING AGGREGATIONS OF HUNDREDS or sometimes thousands occur nearshore along the southernmost coast of South Africa every November and December. The mating system is dramatic and complex. The mating arenas are large and somewhat structured with a mating zone a few feet from the egg clumps in which mating pairs swim and mate. Between there and the egg clumps is the agonistic zone, in which lone large males and smaller sneaker males intercept the mating pairs as they move toward the egg clumps and try to displace the male consort or obtain a quick head-to-head mating. There is intense competition for mates and two distinct mating positions and sperm placement occur in this dynamic reproduction system.

A WELL-REGULATED FISHERY IN SOUTH AFRICA

Fishing Cape Hope squid is a relatively small industry that exports much of the catch to Europe. Most of the squids are captured by small jigging boats. Because of the mobility of these large loliginid squids the annual landings fluctuate a great deal based partly on large-scale changes in the Benguela Current. This species has been studied extensively to better understand these fluctuations and there is a good union between the fishing industry and research academia and government fishery managers. There is a cooperative closed fishery season during November so that sufficient egg laying and recruitment occur for this annual species.

EGG LAYING

A large male is hovering above an egg clump and guarding his female mate, who is inserting an egg finger into the clump. She will then leave the egg clump and the pair will swim together several feet off the substrate to mate, then she will descend again to insert another egg finger, containing 100–300 eggs. This sequence can be repeated a dozen times over several daylight hours.

VEINED SQUID

Loligo forbesii

THIS LARGEST OF THE LOLIGINID SQUIDS has a widespread distribution around Europe but also occurs around the islands of Madeira and the Azores to the west of Portugal. They are preyed upon by pygmy sperm whales, orca (killer) whales, porpoises, seals, and seabirds. Like other loliginids, they lay egg capsules (similar in size to a human finger) in egg clumps that are embedded in the sand in shallow water inshore.

SIGNIFICANT FOR FISHERIES AND BIOMEDICINE

These large squids are fished by bottom trawl in European locations and by jig in the islands of Madeira and the Azores and are a favorite menu item throughout their distribution. They also have the largest nerve fibers known to neuroscience: up to 1.5 mm in diameter. Thus they have been used for many years (mainly in Plymouth, England) as models for how neurons work. This species has been cultured in the laboratory to determine feasibility for making it more widely available to neuroscientists. The growth rate is very fast but full adult size was not attained; moreover, they are active carnivores, thus the challenge of providing live crustaceans and fishes as food render this species too expensive to culture for biomedicine or commercial mariculture.

FAMILY	Loliginidae
OTHER NAMES	Forbes squid, Calamar veteado
TYPICAL HABITAT	Temperate species that lives mostly nearshore on the continental shelf over sand, mud, or seagrass; in the Azores it lives along steep slopes of the islands from 160–3,300 ft (50–1,000 m)
SIZE	Commonly 8–12 in (20–30 cm) mantle length but can be triple that size; males are larger than females
FEEDING HABITS	Feeds on small fishes, crustaceans, and polychaetes; in some locations it feeds on fishes such as cod, whiting, mackerel, and sand eels
KEY BEHAVIORS	Schooling near the bottom or on slopes, sometimes known from telemetry data to "glide" on upwelling currents

DEEP-SEA EUROPEAN SQUID
This large and powerful loliginid is a schooling species that is fished commercially in many locations around Europe and tends to prefer deeper waters than others in this family of squids. Note the streaks of "flame markings" along the ventral mantle.

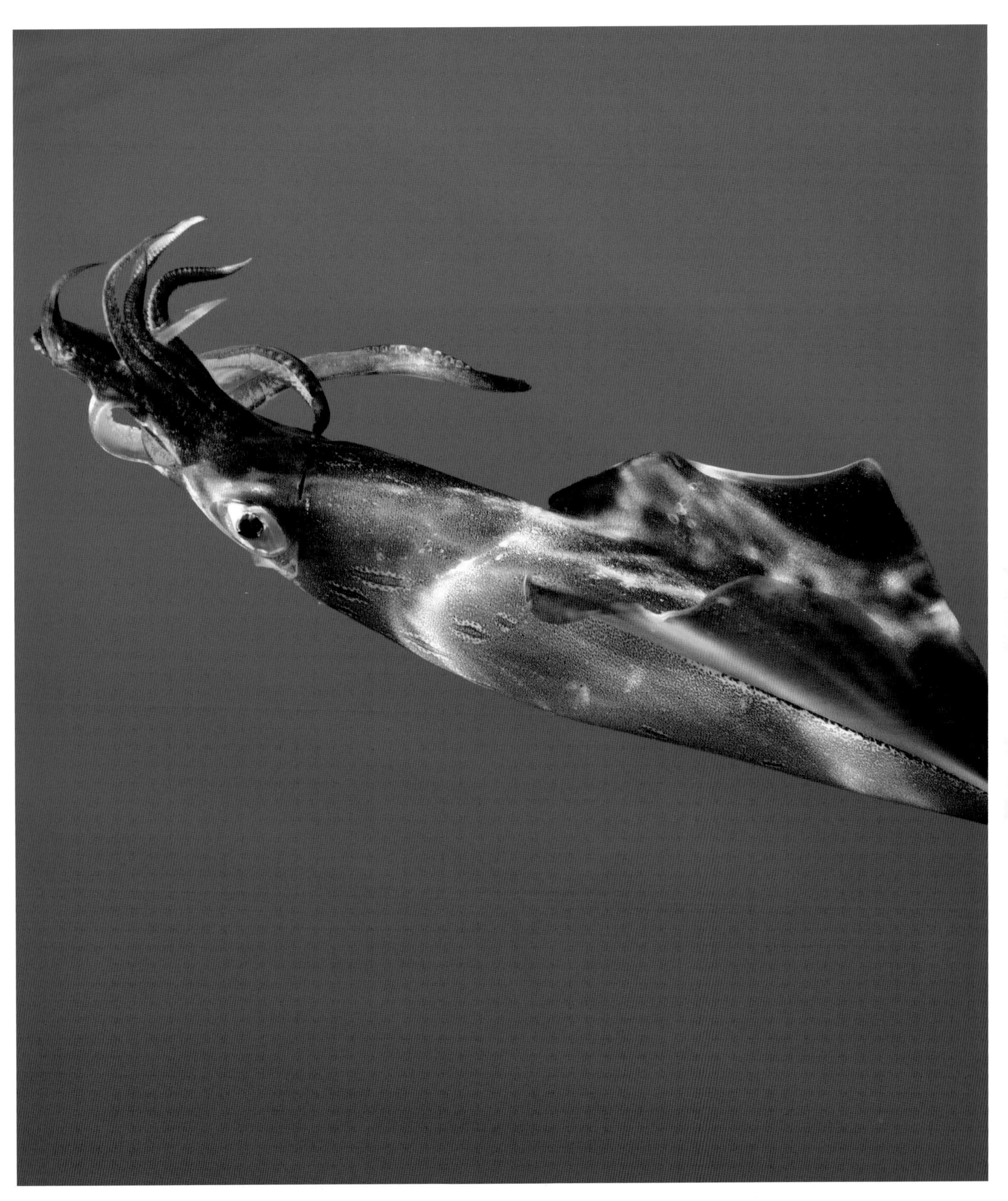

JEWEL SQUID

Stigmatoteuthis arcturi

FAMILY	Histioteuthidae
OTHER NAMES	Strawberry squid
TYPICAL HABITAT	Surface waters to 3,300 ft (1,000 m) depth in the open ocean
SIZE	Mantle length to about 8 in (20 cm); males slightly smaller
FEEDING HABITS	Unknown but probably a mix of small crustaceans and fishes
KEY BEHAVIORS	Bioluminescence; strange body posture to enable it to view upward with one eye and laterally and downward with the other

THE JEWEL SQUID—SO CALLED BECAUSE of the bioluminescent photophores that cover its body—occurs in tropical and subtropical waters of the North and South Atlantic. Their feeding habits are unknown but they are prey to many major predators such as sperm whales and other odontocete cetaceans, various seabirds, sharks, tunas, and lancetfishes. The vertical posture of the jewel squid is rather unusual but an adaptation that allows it to simultaneously look up, down, and laterally. Very little is known about this particular species but the other closely related histioteuthids are all widely distributed in the world's oceans and they all have unusual eyes.

ASYMMETRIC EYES

All squid species in the family Histioteuthidae have markedly asymmetric eyes. The right eye is reasonably "normal" but the left eye is much larger and tubular in shape, with the lens (and thus the eye) pointing at a distinct angle toward the surface. It is thought that this allows the squid to spot prey or predators that are silhouetted against the dimly lit waters above it.

RED GLOW
The many large photophores set in bright red skin are the reason for the common name of the jewel squid; they are sometimes referred to as strawberry squid for the same reason.

JAPANESE FIREFLY SQUID

Watasenia scintillans

THIS VERTICALLY MIGRATING OPEN-OCEAN SQUID is small in size but colorful in two respects: it has an intricate array of luminescent bluish photophores; and it also seems to be able to perceive different colors within the blue and green range. This squid "lives in the dark" by staying very deep during the day (at about 1,000 ft (300 m) where downwelling sunlight is very weak) and migrating to surface waters during the night. In so doing it makes itself vulnerable to attack from below, so its extraocular photoreceptors on the head detect the amount of downwelling light and then regulate the amount of ventral luminescence it displays to hinder detection or recognition of its silhouette as viewed by predators beneath it.

COLOR VISION

Cephalopods are thought to be colorblind. However, an exception is the firefly squid, which has three visual pigments located in different parts of the retina, each having distinct spectral sensitivities. All are in the blue and green portions of the light spectrum, which likely enables firefly squid to discriminate different shades of blue and green but not other colors. This makes ecological sense since they live in dimly lit habitats both day and night, and sources of light would be bioluminescence (which is in the blue spectrum) and starlight or moonlight that is filtered by many feet of seawater, also leaving mostly blue wavelengths.

FAMILY	Enoploteuthidae
OTHER NAMES	Sparkling enope squid
TYPICAL HABITAT	Oceanic. Lives during the day at the mesopelagic boundary about 1,000 ft (300 m) deep; migrates vertically to near surface each night to feed
SIZE	Mantle length up to 2¾ in (7 cm)
FEEDING HABITS	Copepods during paralarval stage; amphipods, euphasids, and fishes in adult stages
KEY BEHAVIORS	Daily vertical migration of several hundred feet; well- developed bioluminescence. Moves inshore to shallow water to spawn. Only cephalopod with color vision.

LUMINOUS BLUE DISPLAYS

The firefly squid is famous for its sparkling displays on the shores of Toyama Bay, where squids gather to mate and spawn, and currents push them in huge abundance into surface waters. Their bioluminescence, emanating from more than 800 photophores on their body surface, illuminates the water surface in electric blue.

GLOSSARY

arm crown
The circle of appendages around the mouth of a cephalopod, including the many arms of a nautilid or eight to ten arms and modified arms (decapod tentacles or vampire filaments) of coleoids.

associative learning
Learning process in which a new response becomes associated with a particular stimulus; a classic example is Pavlov's dog that was taught to associate the sound of a bell with a food award.

axon
The long slender portion of a nerve cell that conducts nerve impulses known as action potentials; the giant axon of the squid is the largest known axon.

benthic
Living on or in the bottom of aquatic habitats.

biomedical model
An animal (or tissue or cell) that has a particular feature that allows targeted research of a specific subject that is of broad medical interest.

bottom trawl
A fishing net that is pulled behind a boat; it has two doors that push it downward, where it is pulled along the bottom to collect bottom-dwelling organisms.

buoyancy
Overall density of an organism relative to that of the surrounding medium, producing its passive tendency to float or sink.

chemotactile
Sensing chemicals by way of touch (as opposed to sensing via water).

chorion
In cephalopods, the semi-rigid membrane that directly encloses the developing embryo.

chromatophore
The cephalopod chromatophore is a complex neuromuscular organ in which a sac of pigment granules can be expanded to show color, or retracted to hide the color, all within half a second. There are three basic colors—yellow, red, and brown— and thousands or millions of these organs in the skin of each animal.

cirrate octopod
An octopod with fins and cirri (finger-like filaments) from the suborder Cirrata.

cleavage
The repeated division of a fertilized egg, producing a cluster of cells.

cohort analysis
Statistical inference of population characteristics based on determination of groups that appear to be similar in age, e.g., based on length-frequency groups (modal size).

deimatic behavior
Startle or threat behavior used as secondary defense against a predator after camouflage fails and the animal realizes that it has been detected.

disruptive coloration
A camouflage tactic in which the cephalopod produces a body pattern with large-scale components of different shapes, sizes, and orientations so that the animal is no longer recognizable as a cephalopod.

DNA fingerprinting
Standardized use of a specific genetic sequence for identification, e.g., of species. For cephalopods, this is often, but not always, approximately 600 base pairs of the mitochondrial gene cytochrome oxidase I.

embodied cognition
The way in which the body form of a cephalopod can influence (positively or negatively) how the nervous system is organized and how it functions. For example, each of the octopus's eight arms has a large ganglion that acts as a sort of "peripheral mini-brain" that

controls several of the arm's functions without full dependence on the large central brain.

episodic-like memory
An animal's ability to encode and retrieve information about what occurred during an episode, where the episode took place, and when the episode happened.

epithelium
A type of animal tissue made up of densely packed cells that rest on a basement membrane to act as a covering or lining of various body surfaces (e.g., skin) and cavities.

euthanize
Chemical or mechanical intervention to end the life of an animal in a laboratory.

evolutionary convergence
Similarities of appearance and function that have evolved independently in organisms that are not closely related, usually as a result of evolutionary responses to similar problems.

extraocular photoreceptor
Cells or an organ that sense light but are in a different location than the eyes.

genome
The complete set of genetic material in an organism.

in situ
For field biology, refers to observation or experimentation in the organism's natural environment. Note: this differs from the medical definition.

in vitro fertilization
Fertilization of eggs removed in a laboratory from a mature female, using sperm removed from a mature male.

iridophores
Small cells in the skin that reflect light of different colors. They complement the colors of the pigmented chromatophores that lie above them in the skin.

leucophores
Skin cells that reflect white to complement the colors of the chromatophores and iridophores.

mantle
The body wall of cephalopods. It encloses the viscera and its contraction aerates the gills and may be used in locomotion.

nanomaterial
A material having particles or constituents of nanoscale dimensions, or one that is produced by nanotechnology. These particles are on the scale of 1-1000 nanometers (there are 1000 nanometers in a millimeter).

nekton
Animals that swim strongly enough to overcome the movements of currents.

nuchal cartilage
A cartilaginous locking structure in decapod cephalopods, located mid-dorsally just posterior to the head, joining and articulating the dorsal head with the mantle near the tip of the gladius.

observational learning
The process of learning through watching others, retaining the information, and then later replicating the behaviors that were observed.

ontogenic
Pertaining to ontogeny—developmental changes throughout an entire life cycle.

osmotic stress
Physiological cost of responding to changes in salinity.

palps
A pair of muscular flaps; found in some cephalopods at the anus or terminal end of the intestine.

parachute attack
Foraging octopuses will approach a likely food source (e.g. a crab, or even a small coral head that will have shrimp or fish or mussels within) and pounce

on it with the web between its arms spread out to envelop the target completely; the extended web looks a bit like a parachute.

pelagic
Living in or otherwise pertaining to the open waters from just above the bottom to the surface of aquatic habitats.

photophore
An organ that produces light either by chemicals or by symbiotic bacteria. Sometimes called a light organ.

phyla (pl. - singular phylum)
The highest level of taxonomic division in the animal kingdom. There are 35 animal phyla, of which Mollusca is estimated to be the second largest.

polarization of light
Light waves that have been transformed from vibration in multiple planes (unpolarized) to waves in which the vibrations occur in a single plane (polarized).

protean behavior
Erratic unpredictable escape behavior and color pattern flashing used as secondary defense after camouflage fails and deimatic behavior has been displayed briefly.

purse seine
A long fishing net with floats on the surface and chains on the bottom; the net is pulled around into a circle and pursed together to entrap fishes and squids.

rapid adaptive coloration
An animal's ability to change its skin patterning and overall appearance within a second or two for camouflage or communication.

reflectin
Protein family that is enriched in aromatic and sulphur-containing amino acids and used by certain cephalopods to manage and manipulate incident light in their environment.

salinity
The concentration of dissolved salts in water.

senescence
Deterioration of form and function late in the life cycle, especially after spawning, resulting in death.

serial reversal learning
Subjects learn to respond differentially to two stimuli. When the task is fully acquired, reward contingencies are reversed, requiring the subject to relearn the altered associations.

sexual dimorphism
Differences in appearance between males and females of the same species.

sneaker male
A cuttlefish or squid male that is too small to fight a big male to obtain a female mate; instead, it uses one of several "sneaky tactics" to gain access to females, who often accept them as mates.

spatial memory
The process through which animals encode information about their environment to facilitate navigation and recall the location of relevant stimuli such as a home den or a food source.

symbiont
An organism that is very closely associated with another, usually larger, organism. This larger organism is called a host.

symbiosis
A close biological interaction between two organisms.

tentacular club
Expanded area at the end of the modified ventrolateral (tentacular) arms of a decapod cephalopod on which suckers and/or hooks are concentrated.

FURTHER RESOURCES

BOOKS

Borrelli, L., Gherardi, F., & Fiorito, G.
A Catalogue of Body Patterning in Cephalopoda.
(Firenze University Press, 2006).

Boyle, P. R.
Cephalopod Life Cycles. Vol. 1: Species Accounts.
(Academic Press, 1983).

Boyle, P. R.
Cephalopod Life Cycles. Vol. II: Comparative Reviews.
(Academic Press, 1987).

Boyle, P. R. & Rodhouse, P.
Cephalopods. Ecology and Fisheries.
(Blackwell Science, 2005).

Darmaillacq, A. S., Dickel, L., & Mather, J. (Eds.).
Cephalopod Cognition.
(Cambridge University Press, 2014).

Hanlon, R. T., & Messenger, J. B.
Cephalopod Behaviour, 2nd ed.
(Cambridge University Press, 2018).

Jereb, P. & Roper, C. F. E.
Cephalopods of the World, An Annotated and Illustrated Catalogue of Species Known to Date, Volume 1: Chambered Nautiluses and Sepioids
(Food and Agriculture Organization of the United Nations, 2005).

Jereb, P. & Roper, C. F. E.
Cephalopods of the World, An Annotated and Illustrated Catalogue of Species Known to Date, Volume 2: Myopsid and Oegopsid Squids
(Food and Agriculture Organization of the United Nations, 2010).

Jereb, P., Roper, C. F. E., Norman, M. D. & Finn, J.K.
Cephalopods of the World, An Annotated and Illustrated Catalogue of Species Known to Date, Volume 3: Octopods and Vampire Squids
(Food and Agriculture Organization of the United Nations, 2016).

Mather, J. A., Anderson, R. C., & Wood, J. B.
Octopus: The Ocean's Intelligent Invertebrate
(Timber Press, 2010).

Nixon, M., & Young, J. Z.
The Brains and Lives of Cephalopods.
(Clarendon Press, 2003).

Norman, M. D.
Cephalopods: A World Guide.
(ConchBooks, 2000).

Rosa, R., O'Dor, R., & Pierce, G. J. (Eds.).
Advances in squid biology, ecology and fisheries. Part 1 - Myopsid squids.
(Nova Science Publishers, Inc., 2013)

Rosa, R., O'Dor, R., & Pierce, G. J. (Eds.).
Advances in squid biology, ecology and fisheries. Part 2 - Oegopsid squids.
(Nova Science Publishers, Inc., 2013)

Saunders, W. B. & Landman, N. H.
Nautilus: The Biology and Paleobiology of a Living Fossil
(2nd ed.) (Plenum Press, 2009)

Staaf, D.
Squid Empire; The Rise and Fall of the Cephalopods.
(University Press of New England, 2017)

Wells, M. J.
Octopus; Physiology and Behaviour of an Advanced Invertebrate.
(Chapman and Hall, 1978).

WEBSITES

tolweb.org/Cephalopoda

mbari.org/tag/cephalopods

ucmp.berkeley.edu/taxa/inverts/mollusca/cephalopoda.php

INDEX

Bold type indicates pictures.

ACKNOWLEDGMENTS

We are most grateful to Ivy Press for initiating this endeavor. In the early phases we appreciated the professional guidance and patience of Commissioning Editor Kate Shanahan and Editorial Director Tom Kitch. University of Chicago Press Editorial Director Christie Henry was very supportive of this effort and was instrumental in orchestrating the joint publication. As the actual writing and photograph collection proceeded over a 2-year period, we worked most closely with Ivy's Project Editor Joanna Bentley. Her enormous experience combined with her gentle and effective nudging to meet multiple deadlines are deeply appreciated.

Roger thanks many dive partners and lab colleagues over the years who enabled a great deal of discovery of cephalopod biology. Mike appreciates his wife Susan and her decades of tolerance for his long absences at sea that were necessary to learn some of the information on deep-sea cephalopods presented here. Louise is hugely grateful for the support of Felix and Poppy, who let Mum write when she should probably have been playing, and Mark, who made sure nobody starved.

PICTURE CREDITS

The publishers thank the following for permission to reproduce the images in this book and for use of reference material for illustrations. All reasonable efforts have been made to contact copyright holders and to obtain their permission for the use of copyright material. We apologize for any unintentional omissions and will be pleased to incorporate any corrections in future reprints.

Hideki Abe 137t
Alamy Stock Photo / F1online digitale Bildagentur GmbH 1,26, /Andrey Nekrasov 2,50, /WaterFrame 7b, 97t, 115t, 116b, 123t, /Nature Picture Library 18-19, 75, /Mark Conlin 24r, 192, 201, /David Fleetham 26l, /Juniors Bildarchiv GmbH 39bl, /Visual&Written SL 41l, /imageBROKER 54-55, /Papilio 65t, /Pete Niesen 104-105, /Michael Greenfelder 125l, /Arco Images GmbH 127r, /Jane Gould 135t, /Wildestanimal 152-153, /Juergen Freund 161b, /Minden Pictures 171, /RGB Ventures/SuperStock 177, /National Geographic Creative 188-189, JTB MEDIA CREATION, Inc. 191, /Ed Brown Wildlife 198t, /Michael Patrick O'Neill 206t, /Reinhard Dirscherl 213
Ardea / © Valerie & Ron Taylor 141, /© Auscape 142t, /© Gabriel Barathieu/Biosphoto 158
Azote / Mattias Ormestad 197
Gregory J. Barord 86
California Department of Fish and Wildlife 120t
Kay Cooper 194
Dreamstime.com / © Serban Enache 193t
Alessandro Falleni 98t
Danté Fenolio 65c
Eliot Ferguson 108
FLPA / Photo Researchers 3, 7t, 217, /Fred Bavendam /Minden Pictures 42, 70, 76, 129r, /Reinhard Dirscherl 58, /Norbert Wu/Minden Pictures 85t, 113c, /Colin Marshall 92b, /Richard Herrmann/Minden Pictures 100
Dirk Fuchs 56c, 56b, 67, 72
Roger Furze 60r
Getty Images / Hal Beral 6, /Brian J. Skerry 9, 78t, /Dave Fleetham 14b, 124, /Mark MacEwen/naturepl.com 15, /Andrey Nekrasov 16l, Carrie Vonderhaar /Ocean Futures Society 17b, 20, /Henrik Sorense 22tl, /Louise Murray/robertharding 22bl, 138, /Wild Horizons/UIG 26, 37br, 40, 154, /Franco Banfi 27, 190, /Digital Camera Magazine 30t, /Jeff Rotman 36, /Marevision 37bl, / _548901005677 38, /Andy Reisinger/ Hedgehog House/Minden Pictures 41r, /Ingo Arndt/Minden Pictures 60l, /Colin Keates 62r, /Visuals Unlimited, Inc./Michael Ready 77b, /Jennifer Hayes 78b, /Daniela Dirscherl 79, /Kike Calvo 113r, /Rodger Klein 114, /Jean Lecomte 119t, 122r, /David Fleetham/Visuals Unlimited, Inc. 121t, /De Agostini Picture Library 129l, /Andrey Nekrasov 135b, /Image Created by James van den Broek 142b, /Reinhard Dirscherl 155r, 175, 182b, /Georgie Holland 169t, /Bob Chamberlin 193bl, /Luis Javier Sandoval 196, /Nature, underwater and art photos. www.Narchuk.com 199, /Science VU 200
Stephen J. Hamedl 89t
Richard Hamilton 59
Roger Hanlon 14t, 16r, 25t, 25cl, 94, 107b, 109, 144, 157, 161t, 162, 163tl & b, 164, 165, 166, 167, 169b, 172, 181, 182t, 185, 186, 205, 210t
JMG / Image Quest Marine 202
iStock / demarfa 8, /Miya0105 51t, /4kodiak 190b
Tom Kleindinst, Woods Hole Oceanographic Institution 195
T. Kubodera, National Museum of Nature and Science, Japan 82
Minette Layne 39br
Nathan Litjens 31
Tomas Lundälv 130r
MBARI / © 2005 MBARI 65b, /© 2001 MBARI 66l, /© 2009 MBARI 90, /© 2002 MBARI 115b, 122l, / ©2012 MBARI 133r /© 2004 MBARI 137b
Ryo Minemizo Photography 68b
© Museo di Storia Naturale di Milano 62l
NOAA / Marine Flora and Fauna of the Eastern United States. Mollusca: Cephalopoda 37t / NOAA Fisheries 49, /Image courtesy of the NOAA Office of Ocean Exploration and Research, Discovering the Deep: Exploring Remote Pacific MPAs 113l, /NOAA OKEANOS Explorer Program, 2013 Northeast U. S. Canyons Expedition 125r, 148t
Nature Picture Library / Georgette Douwma 10, /David Shale 21cr, 21br, 22cl, 25c, 66r, 68t, 126r, 131r, 132, 147, 151, /Nature Production 23, 116t, 118r, /David Fleetham 25cr, 85b, 126l, /Alex Mustard 31b, 34, 163tr, 206b, /Peter Scoones 46, /Doug Perrine 103, /Sue Daly 112, /Juergen Freund 127l, /Visuals Unlimited 130l, /Solvin Zankl 131l, 136
Steve O'Shea 71
OceanwideImages.com /Rudie Kuiter 118l
Oxford University Press / Young, J. Z. The Anatomy of the Nervous System of Octopus Vulgaris (1971) 39t
Photoshot / © NHPA 52
Press Office Münster / Angelika Klauser 81r
Mark A. Royer 208
SeaPics.com / © Doug Perrine
Alexander Semenov 28
Science Photo Library / Pascal Goetgheluck 29bl, 29br, /Reinhard Dirscherl 107t, /Dante Fenolio 111tr, 214, /Andrew J. Martinez 120
Dan Seddon 77t, 92t
Shutterstock / Nonthawit Doungsodsri 4, 61, /Dongseun Yang 30c, Boris Pamikov 45, Kondratuk Aleksei 80, Richard Whitcombe 155l, 192br
Adapted by permission from **Springer Nature:** Experientia, General morphological and functional characteristics of the cephalopod circulatory system. An introduction, R. Schipp, Copyright © 1987 33
Stefan Siebert 119b
Barry Sutton 84
Tanabe, Hikida and Iba, Two Coleoid Jaws from the Upper Cretaceous of Hokkaido, Japan. Journal of Paleontology, 80(1), 2006, pp. 138–145 81b
Nick Terry 73t
Ilan Ben Tov 81l
Undersea Hunter/DeepSee Submersible 69
K.G. Jebsen Centre for Deep Sea Research, University of Bergen 133l
Michael Vecchione 21t, 111br, 178
Richard E Young 32, 35, 156 (modified from Cloney, R. A. and E. Florey. Ultrastructure of cephalopod chromatophore organs. 1968. Zeits. für Zellforsch. 89:250-280)
Cunjiang Yu 198b